REPRODUCTION IN CATTLE

Edited by
Cecilia Umali

ArclerPress

Reproduction in Cattle

Edited by Cecilia Umali

Edition 2018

ArclerPress

Published by
Arcler Press LLC
708 3rd Avenue, 6th Floor
New York
NY 10017
United States of America
www.arclerpress.com

ISBN: 978-1-68094-593-5 (Hardback)

Printed in United States of America

Library of Congress Control Number: 2017933194

About the Editor

Cecilia Umali

Cecilia finished Bachelor of Science in Agriculture in 1984 at the University of the Philippines. She specialized in Dairy Technology. Her interest is in biosynthesis of flavor compounds by microorganisms. Her thesis Manufacture and Characteristics of Camembert Cheese from Goat's Milk was aimed at masking the goat flavor with flavors synthesized by Camembert microorganisms. She serves as Food Consultant to small and medium enterprises.

Table of Contents

2 CATTLE NUTRITION

3 CATTLE HEALTH

7 A.I. TECHNIQUE IN CATTLE

8 PREGNANCY DIAGNOSIS

9 LIVESTOCK PRODUCTION SYSTEMS

List of Abbreviations

FAO	Food and Agricultural Organization
FCM	Fat Corrected Milk
HCl	Hydrochloric Acid
ICCC	Individual Cow Cell Counts
NSIP	National Sheep Improvement Program
SIT	Sterile Insect Technique
WHO	World Health Organization

Preface

Cattle are the most common type of large domesticated ungulates. Cattle are raised as livestock for meat (beef and veal), as dairy animals for milk and other dairy products, and as draft animals (oxen or bullocks that pull carts, plows and other implements). Other products include leather and dung for manure or fuel. They are an integral part of animal agriculture in all parts of the world. Demand for some products, such as fat in meat and in milk, has declined in some markets, and in some countries milk production is limited by quotas. Nevertheless, the efficient production of animal protein continues to be of vital importance. Efficient production system can maximize output of meat and milk per unit of feed input per animal. In cattle production as with other livestock systems a high level of reproductive performance is absolutely crucial to efficient production. Efficient reproduction results in optimum calving intervals, which in turn result in optimum production of milk and calves to be reared for beef or replacements. The breeding program may use either artificial insemination (AI) or natural service. AI has been commercially available; it is widely used in dairy cattle but is used much less in beef cattle because of handling and labor costs. AI offers a selection of bulls with known genetic potential, such as measured by estimated breeding values (EBVs) for traits such as ease of calving or growth rates. When nutrition and heat detection are properly managed, satisfactory results are obtained. Failure to detect estrus is the major reason for unsuccessful AI. When cows are properly inseminated with good-quality semen at the proper time, 50%–60% may conceive on first service, the same percentage on second service. In cattle, AI is used primarily for genetic improvement of livestock and to facilitate high health replacement strategies. The worldwide adoption of AI for genetic improvement in dairy cattle was made possible by development of a progeny test system and subsequent use of milk production records as an objective measure of performance on which to select superior bulls, techniques for freezing semen, and liquid nitrogen storage refrigerators. The development of objective systems such as EBVs to measure economic traits in beef cattle (e.g., growth rate, carcass conformation and composition, efficiency of feed conversion) and thus the more accurate selection of sires, as well as control of the estrous cycle, is leading to an increase in use of AI in beef cattle. Processing of frozen semen is a highly specialized technique. In order to manage cattle reproduction and the specific problems that arise, it is important to understand the underlying physiology of the reproductive process.

Content Coverage

Chapter One presents a general overview of reproduction management. Modern techniques applied to reproduction of domestic mammals aim at increasing the efficiency of production of offspring under the conditions of management best suited to the farmers.

Chapter Two is about animal nutrition, which is important for a variety of reasons. Animals need the proper nutrition for growth and maintenance, and to provide energy for work and vital functions. Maintenance is the nutrition required for an animal to maintain its current weight. Energy is the ability of the body to perform functions. Proper nutrition is also needed to maintain body temperature, produce milk, reproduce, and develop proper bone structures.

Chapter Three focuses on maintaining animal health, which is the key to a profitable and satisfying animal enterprise. There are several considerations that need to be made in dealing with the health of animals.

Chapter Four presents an emphasis on reproductive anatomy and physiology of cattle. In order to understand why an animal displays the many signs of estrus, when she should be inseminated, and how the pregnancy develops, you must clearly understand the hormonal mechanisms controlling the estrous cycle in cattle.

Chapter Five focuses on breeding and breeding techniques. The more modern techniques involve a wide variety of laboratory methods, including the modification of embryos, sex selection, and genetic engineering. These procedures are beginning to supplant traditional breeding methods, which focus on selectively combining and isolating livestock strains.

Chapter Six highlights about the basics of heat detection. Efficient and profitable reproductive performance of a dairy herd requires routine but conscientious heat detection and proper timing of artificial insemination. Failure to detect estrus (heat) is a major factor contributing to low fertility.

Chapter Seven focuses on artificial insemination techniques. The recto-vaginal technique is the most commonly used method to artificially inseminate cattle. The basic skills required to perform this technique can be obtained with about three days practice under professional instruction and supervision.

Chapter Eight is focused on the accurate diagnosis of pregnancy, which is of crucial importance in the establishment and maintenance of optimal reproductive performance. It is desirable for the farmer to know as soon as possible if a mated cow is not pregnant, so that she can be rebred with the minimum delay.

Chapter Nine deals with the concept of livestock production systems. The majority of livestock kept by smallholder farmers are integral components of complex crop-livestock systems. Building a complete understanding of the biological, social and economic aspects of these systems is a necessary ingredient for determining intervention points and the development of appropriate technologies for smallholder farmers.

CHAPTER 1

Reproduction Management: General Overview

Chapter Contents

- **Introduction**
- **Bull Reproductive Management**
- **Pregnancy Determination in Cattle**
- **Methods of Animal Breeding**
- **Breeding of Cattle and Buffaloes**

Modern techniques applied to reproduction of domestic mammals aim at increasing the efficiency of production of offspring under the conditions of management best suited to the farmers. This report presents the different techniques now available to achieve this objective. **For the male**, in addition to the semen in artificial insemination now possible in all species of farm animals, emphasis will first be on the possibility to deliver a small number of spermatozoa from the best pedigree sires to a maximum number of females with the best chance of fertilization and, second, to keep the males of seasonal breeders (like sheep and goats) permanently at the top of their potential sperm production.

INTRODUCTION

Dairy and beef producers should strive to increase reproductive efficiency as a key driver of economic efficiency in the sector. Reproductive efficiency, or "pregnancy rate," is defined as the proportion of cows eligible to be bred that become pregnant during an estrous cycle (or approximately 21 days), and which determines the calving to conception interval at the end of the voluntary waiting period. As pregnancy rate increases in dairy herds, the calving to conception interval decreases, and the herd status becomes, on average, less "days in milk" (DIM). This has the effect of increasing the potential amount of milk produced per day of herd lifetime, because yield classically declines at 0.3% per day after peak lactation production. A major and realistic goal of every beef cow/calf operator should be to raise or market 85 calves per 100 cows every year. Greater reproductive efficiency also reduces the number of cows culled for reproductive failure; collectively, these changes increase herd income.

Reproductive performance in both beef cow/calf and dairy operations can be improved by the following: 1) properly identifying and managing animals to carry out reproductive programs; 2) keeping records that enable determination of important herd indices, such as percent calf crop, pregnancy rate, length of calving season, culling rates, calf morbidity and mortality, breeding efficiency of bulls, and performance and production information; 3) meeting the nutritional requirements of various classes of livestock in the herd, emphasizing nutritional needs and cost efficiencies; 4) establishing a breeding program for heifer replacements and cows; 5) practicing sire selection and reproductive management; 6) adopting a vaccination/immunization program for the cow/calf herd, bulls, and calves; 7) evaluating reproductive failure and abortions; 8) providing adequate facilities; and 9) ensuring that the calf is well cared for at birth and receives adequate colostrum.

Nutrition in Cattle Reproduction

Nutrition is one of the most important management factors in reaching calf crop goals and in attaining a short calving season every year in beef breeding herds. The limiting nutrient related to reproduction in beef and dairy cattle is usually energy; although dairy cattle are usually fed rations that supply adequate energy during lactation, genetic drivers for milk production inevitably lead to a period of some negative energy balance postpartum. The level of energy and body condition before calving primarily influences when a beef cow returns to estrus, whereas level of energy after calving primarily influences subsequent conception. Feed requirements vary during the reproductive cycle (see Nutritional Requirements of Beef Cattle).

There are four periods of beef cow nutrient requirements, and generally three for dairy cows. **Period 1** is the interval from calving to breeding; it is ~70–90 days and is the period of greatest nutritional demand. The dairy cow is at maximal milk production and recovering from the stress of parturition. During this period, she is expected to be ready to breed.

Period 2 is the interval from rebreeding to weaning the beef calf; it is ~120–150 days in beef cows. Periods 2 and 3 overlap in dairy cows and are not as easily separated as in beef cattle. The beef cow should gain weight while still milking. Although some dairy cows maintain body weight, many high producers continue to lose weight during this period.

Period 3 is from weaning to 50 days before calving; it lasts ~100 days and is the period of least nutritional demand. The beef cow has only to maintain her condition and continue fetal development. The dairy cow should be managed to gain or lose body weight during the last months of lactation to achieve target body condition ready to enter the stable dry period.

Period 4 is a critical stage and is the 50 days before calving; it is during this time that 75% of fetal growth occurs. Cows are usually not lactating during this "dry period." Cow condition at calving is critical to rebreeding; the onset of estrus after calving is delayed in cows that lose weight or are thin and not gaining during late pregnancy.

Dairy cows (see Nutritional Requirements of Dairy Cattle) are usually fed for optimal milk production throughout their 305-day lactation. It is assumed they will lose weight during heavy lactation (the early months) and regain the loss during the remainder of lactation. Dairy cows should not be overfed during the dry period because of a genetic predisposition to sacrifice body condition to maximize milk production through insulin resistance. This leads to the increased probability of metabolic diseases, eg, type II ketosis and fatty liver disease (see Ketosis in Cattle and see Fatty Liver Disease of Cattle), during early lactation as insulin resistance leads to excessive fat mobilization from adipose tissue storage and overwhelms the capacity of lipoprotein transport mechanisms in the bovine liver to transport and metabolize lipid. In addition, dairy cows should be fed to minimize the incidence of calving-related disorders (eg, dystocia, hypocalcemia, and retained fetal membranes), including control of dietary cation-anion balance (DCAB), which have a negative effect on fertility and health postpartum.

The amount of cow feed required per pound or kg of calf weaned is fairly constant, although larger cows require more feed for maintenance than smaller cows. Cows that give more milk require more feed, generally with a higher level of protein. Increased milk is produced at the expense of reproduction when feed is not adequate to meet all needs.

The protein requirement of young growing stock and heavy-milking cows is often a limiting factor, while mature dry cows are often overfed protein. Heifers must be fed adequately from weaning to breeding if they are to calve at 2 yr of age; this target is critical for herd economics, because before this point the absence of beef calf or milk for sale represents significant investment and risk.

To provide the essential nutrient requirements during various stages of the reproductive cycle, major forages and homegrown cereals should be analyzed to monitor nutrient content and actual value. Variation in amounts of trace minerals is common between and within different geographic areas. Globally, different systems are used to determine energy levels of the ration, such as the metabolizable protein (MP) or «Feed into Milk» (FiM) models or the Total Digestible Nutrient System and the California Net Energy System. All are commonly used, and application should be tailored to fit the individual operation.

Even within nutritional need categories, cattle benefit from feeding and handling in subgroups: lightweight heifers at weaning need to gain more than heavier heifers to reach puberty by breeding season; first-calf heifers require special attention from both an energy and competition standpoint if they are expected to breed and conceive at the proper time. These heifers are still growing, as well as lactating, and they may not have the rumen capacity to meet postcalving energy needs on forage alone. Monitoring of growth rates is important to achieve successful rearing targets. Supplemental feeding of both high-energy and high-protein feeds to first-calf heifers may be required for optimal reproductive potential. Calves from first-calf beef heifers may be weaned 30–40 days earlier than calves from cows in the main herd to allow the heifer more time to grow and recover from demands associated with lactation.

Thin, old, and small cows may not compete favorably with heavier cows within the same herd and often benefit from being fed as a separate subgroup.

Lactating dairy cattle are usually fed according to milk production. They may be fed concentrate on an individual basis or divided into groups according to milk production and fed an appropriate total mixed ration.

Breeding Program for Heifer Replacements and Cows

If a cow is to calve consistently, she must deliver her first calf early. Puberty is a function of breed, age, and weight. Beef heifers that are bred at 13–15 mo and calve at 22–24 mo have two advantages: they get closer attention from herdspeople by calving before the main herd starts to calve, and subsequently they have the extra time needed to rebreed with the mature cow herd. For heifers to breed at 14 mo, they should have attained at least 65%–75% of their projected mature weight; therefore, adequate nutrition is of major importance. The breeding season for virgin beef heifers should start 3 wk before that of the main cow herd. The above considerations do not apply to dairy cattle, which calve throughout the year; however, scheduling heifer calvings at the start of a seasonal calving dairy herd such as in New Zealand or Ireland represents an opportunity to "reset" the calving pattern. Lifetime profit of dairy replacement heifers is maximized when heifers calve at 23–25 mo of age. Thus, to maintain genetic progress and maximize profitability, heifer breeding strategies on dairy operations should include artificial insemination (AI) that results in attainment of pregnancy to allow calving at ~24 mo of age.

To compensate for the greater attrition rate usually seen with virgin heifers, a greater number should be bred than is needed to maintain or increase herd numbers, eg, 150%.

Irregularities of Estrus and Anestrus

Breeding will not occur if the cow is anestrous, or if estrus is undetected. Genuine anestrus is not common in dairy cows but is more common in postpartum beef cows in below target body condition. Anestrus, subestrus, or irregular estrous cycles in cows may result from a number of factors, including poor management or nutrition, disease, injury, or disturbances in endocrine functions. One of the most important management factors in artificially bred herds is failure to detect or observe estrus. The average duration of estrus is 18 hr, but in many cows it is appreciably shorter. A systematic program for detection of estrus is important if cows are to be bred at the right time. A producer must be familiar with signs of estrus. Aids in estrus detection that are valuable adjuncts to the heat-detection program include chalk marks on the tailhead, chemically or electronically activated devices attached to the tailhead of the cow that reveal when other cows have mounted, and a vaginal probe that measures the electrical conductivity of the vaginal mucus. Accidental access of bulls to cows and failure to keep proper breeding records often result in apparent anestrus because of pregnancy without a service history.

Although genuine anestrus is generally not common in dairy cows, "subestrus" with poor estrous expression or irregular estrous cycles in cows may result from a number of factors, including poor management or nutrition, disease, injury, or disturbances in endocrine functions. In many dairy herds, detection of estrus for AI is inefficient, because not all cows are identified in estrus because of human error, attenuated expression of estrus in high-producing cows, and adverse responses to heat stress. Systematic breeding programs for AI at a predetermined time (ie, timed AI [TAI]), without the need for estrus detection, coupled with early rebreeding of nonpregnant cows are successful options for reproductive management of lactating dairy cows. These systems optimize pregnancy rate by synchronizing follicle development, regression of the corpus luteum (CL), and precise induction of ovulation to provide a fixed TAI. Incorporation of TAI in dairy herd reproductive management programs reduces labor requirements for detection of estrus while improving overall reproductive performance and maximizing profit.

Silent heat refers to normal follicular development and ovulation without evident signs of estrus. Its frequency decreases as lactation progresses, so that incidence is low by 4 mo postpartum. Cows with true silent heats may be detected only through rectal palpation or ultrasound of the ovaries, via accelerometers/activity meters, or the use of progesterone assay in milk or plasma.

Cyclic changes occur in the ovary over 18–24 days. These changes generally can be recognized and the time of the cycle estimated, particularly in the 3–4 days before ovulation, at the time of ovulation, or 3–4 days after ovulation. The CL regresses 3–4 days before the onset of estrus; it becomes smaller in size and changes from a diestrous, liver-like consistency to one more fibrous. Estrus is evidenced by the presence of a palpable follicle, an absent or regressed CL, and firm uterine tone. The vaginal mucosa is edematous, the cervix is relaxed and hyperemic, and a variable amount of clear serous mucus is frequently seen at the vulva, which is puffy and swollen. The immediate postovulatory period is characterized by blood in the mucous discharge and an ovary with a corpus hemorrhagicum, which on palpation is recognized as a soft area (5–15 mm in diameter) in the ovary. The CL is detectable by day 4–5 as a small and somewhat softer structure than the mature CL, which reaches maximal size by day 7.

The examiner can sometimes predict the next estrus from previous observations around estrus, and the cow can be watched closely at the next anticipated estrus. The primary behavioral sign of estrus—standing to be mounted—is the accurate predictor of ovulation. In cows that are approaching ovulation, the appropriate time can be estimated, and the cow bred on secondary signs such as mounting other cows; however, breeding on secondary signs may be associated with wide variation in time to ovulation. Aged gametes (either semen or oocytes) are associated with reduced embryo viability.

Regimens have therefore been developed for administration of prostaglandins (PGs) and their analogues alongside other products to synchronize estrus and to reduce dependence on estrus detection. Prostaglandins are effective only if a cow has a functional CL. A range of synchronization options are in common use, including double prostaglandin regimens, GnRH and prostaglandin protocols, and regimens that use combinations of GnRH and prostaglandin alongside intravaginal progesterone-releasing devices. For estrus synchronization, the prostaglandin or its analogue is administered to all cows. In those in days 6–18 of the cycle, the CL will regress and estrus will occur in 2–7 days. The others may either have been in estrus recently or will be in a few days. Eleven days later, all cows will be between days 6 and 18 of their cycle, and prostaglandin is administered a second time. Most cows will be in estrus in 3–4 days and will ovulate in 4–5 days. Breeding is done either on signs of estrus or fixed time. Cystic ovary disease (see Cystic Ovary Disease) may be responsible for irregularities of the estrous cycle, eg, follicular cysts (anestrus, nymphomania, and shortened cycles) and luteal cysts (anestrus).

Under certain circumstances, ovaries are nonfunctional. They can be recognized as smooth, small, bean-shaped structures on a single examination, or reveal no activity or change after several examinations over a period of 3 wk. The most common causes are low total energy intake during late winter or droughty summer pastures, and excessive loss of body weight in lactating dairy cows postpartum.

The stress of chronic or severe disease, injury, or ovarian tumors may interrupt ovarian activity and result in anestrus. Congenital defects, such as freemartinism and ovarian hypoplasia, result in estrual failure. Inactive ovaries are treated by correcting the basic cause; they usually do not respond to gonadotropin or steroid hormone treatment.

BULL REPRODUCTIVE MANAGEMENT

A desirable goal for beef producers is a 95% calf crop delivered within 45–65 days, with an optimal weaning weight obtained at the most efficient cost. Bull selection, management, and evaluation of performance are integral aspects of beef improvement. The bull can affect calving percentage as well as quality of calves. The use of performance-tested bulls (beef and dairy) is recommended for both natural breeding and artificial insemination.

A disease control program for bulls should include the following procedures: 1) Before use, bulls should be checked for brucellosis, tuberculosis, trichomoniasis, and paratuberculosis (ie, from a herd free of paratuberculosis). Bulls previously used in other herds, particularly herds in which disease status is not known, may spread diseases, particularly campylobacteriosis and trichomoniasis. Prophylactic sheathwashing with antibiotics may reduce the risk of *Campylobacter* transmission. 2) Depending on prevailing health status and market requirements, bulls should be vaccinated against infectious bovine rhinotracheitis, bovine viral diarrhea, clostridia, *Haemophilus*, campylobacteriosis, and leptospirosis if available. Vaccines should be administered around puberty, at 6 mo and 1 yr of age, and then annually, 1 mo before the breeding season. 3) Breeding soundness examinations should be performed annually at the most economically advantageous time for the producer (usually 1 mo before the breeding season).

All bulls should be on the premises 2 mo before the breeding season to allow them to adapt to the environment. Isolation of all new additions to the herd (bulls or cows) is recommended for proper adaptation and preparation. A breeding soundness examination consisting of a thorough physical examination, including internal and external genitalia, measurement of scrotal circumference, and microscopic examination of semen for sperm motility and morphology, should be conducted ~1 mo before the breeding season. Opportunities are emerging for more objective methods of semen evaluation to be used alongside standard microscopy, such as computer-assisted semen analysis (CASA) for motility and flow cytometry to assess morphologic defects. (Also see Breeding Soundness Examination of Bulls.)

During the breeding season, the bull should be watched closely for mating behavior. The standard recommendation is 25 cows per bull. There are significant variations in this ratio, depending on the breeding soundness and libido of individual bulls, differences in terrain, and length of breeding season.

The weight loss that develops in bulls during the breeding season should be restored during the off-period, but overconditioning should be avoided.

Breeding in Cattle Reproduction

The breeding program may use either artificial insemination (AI) or natural service. AI has been commercially available for >60 yr; it is widely used in dairy cattle but is used much less in beef cattle because of handling and labor costs. AI offers a selection of bulls with known genetic potential, such as measured by estimated breeding values (EBVs) for traits such as ease of calving or growth rates. When nutrition and heat detection are properly managed, satisfactory results are obtained. Failure to detect estrus is the major reason for unsuccessful AI. When cows are properly inseminated with good-quality semen at the proper time, 50%–60% may conceive on first service, the same percentage on second service.

Embryo transfer (see Embryo Transfer in Farm Animals) is frequently used to increase the number of progeny from the most valuable beef and dairy cows. Sexing of semen has been adapted to commercial field usage and is increasingly used. Sexing of embryos is becoming more readily available and practical for field usage, and some cloning techniques are emerging in availability.

Heifers should be bred according to size and age at puberty; at first breeding they should be 65%–70% of their projected mature body weight, with dairy heifers achieving >125 cm withers height at this stage. Selection of bulls to be used for natural service should be based on likely size of the calf at birth; the bull's own birth weight (not his adult weight) is a useful guide, but genetic proofs such as EBVs are vital.

Heat synchronization of heifers and cows is possible, but such programs depend on adequate management and cooperation. Also, sufficient skilled labor to breed and assist during calving is essential.

Artificial Insemination (AI)

In cattle, AI is used primarily for genetic improvement of livestock and to facilitate high health replacement strategies. The worldwide adoption of AI for genetic improvement in dairy cattle was made possible by development of a progeny test system and subsequent use of milk production records as an objective measure of performance on which to select superior bulls, techniques for freezing semen, and liquid nitrogen storage refrigerators.

The development of objective systems such as EBVs to measure economic traits in beef cattle (eg, growth rate, carcass conformation and composition, efficiency of feed conversion) and thus the more accurate selection of sires, as well as control of the estrous cycle, is leading to an increase in use of AI in beef cattle.

Processing of frozen semen is a highly specialized technique. Attention to detail at each step is important to maintain semen quality. The freezability of semen varies among bulls. However, semen of high motility and morphology quality generally freezes well. Best results are obtained when semen is processed in a properly equipped laboratory by experienced staff at an AI center.

Collection and Handling of the Semen Sample:

(Also see Breeding Soundness Examination of the Male.) Semen is collected using an artificial vagina or by electroejaculation (electrical stimulation of the seminal vesicles and ampullae). As long as the sample is of high quality, freezability and fertility should be normal. These techniques should not be used if the bull is unable to naturally service a cow for reasons that could be genetic.

Most AI in cattle today is performed with frozen semen. Frozen semen may be maintained for years; extenders permit more insemination doses to be processed from one collection of semen, maintain the fertility of the semen longer, protect the spermatozoa from sudden temperature or pH change, and prolong viability. Semen is usually extended with citrate-buffered egg yolk or heat-treated skim milk plus glycerol, sugars, enzymes, and antibiotics. Final extension is designed to package 0.25 mL or 0.5 mL of semen containing 20–30 million spermatozoa at time of freezing.

Extenders are often divided into fraction A and fraction B. The initial extension of semen is done with fraction A at the same temperature, eg, 86°F (30°C). The extended semen is then cooled to 41°F (5°C) over 40–50 min, or more slowly. Holding the extended semen at this temperature for 3–4 hr enables the antibiotics in fraction A to complete their action before being inhibited by the cryoprotectant glycerol. Fraction B contains a cryoprotectant such as ethylene glycol or glycerol (eg, 14%) and is added at 5°C in equal quantity to the extended semen. Each AI center has its own standard extenders and processing procedures. Glycerol (11%–13%) may be used with milk-based diluents. Before freezing, semen should be stored for 4–18 hr at 5°C.

For freezing, bull semen is usually packaged in appropriately identified plastic straws (0.25 or 0.5 mL). Optimal freezing rates are known for many cell types, and spermatozoa can withstand a wide range of rates. In practice, extended semen is frozen in liquid nitrogen vapor before being plunged into liquid nitrogen at –320°F (–196°C). Storage in liquid nitrogen tanks is safe for ≥20 yr, and semen is transported in such tanks. The level of liquid nitrogen in tanks must be monitored to avoid semen losses, which is seen when the tanks become defective or when liquid nitrogen gradually evaporates.

Because spermatozoa do not survive for long after thawing, the semen should be used immediately. Thawing is best done as quickly as possible without damaging the semen by overheating. In practice, straws may be thawed in warm water (95°–98°F [35°–36.5°C]) for ≥30 sec and immediately placed in the cow's reproductive tract. Recommendations by the AI center that processed the semen should be followed.

Insemination Technique

The rectovaginal method is used almost exclusively. After thoroughly cleaning the external genitalia with disposable toweling, one gloved hand is introduced into the rectum and grasps the cervix. The insemination pipette is introduced through the vulva and vagina to the external cervical os. By manipulating the cervix, along with light cranial pressure on the pipette, the pipette is advanced through the annular rings of the cervix to the junction of the internal cervical os and the body of the uterus. The semen should be expelled slowly (5 sec) to avoid sperm loss. If insemination records and consistency of the cervical mucus suggest possible pregnancy, the pipette should be advanced less than one-half of the way through the cervix, and the semen expelled. The optimal time to inseminate is between the last half of standing estrus and 6 hr thereafter, which is described as the "am/pm rule," in that cows observed beginning standing heat in the morning should be inseminated that afternoon, etc.

If fertility problems arise when AI is being used, the semen should be investigated, although many factors other than semen are involved in attaining high fertility. Motility after thawing is an important criterion. An adequate number of motile spermatozoa at the time of insemination is critical. Morphologic examination also helps assess the role of semen in infertility cases. Comparisons within herds of diagnosable pregnancies resulting from the suspect semen and from semen from other bulls may be useful. Estrus detection continues to be the most important factor that influences AI efficiency. This factor should be investigated first, and inseminator proficiency second. The latter includes an evaluation of thawing temperature, time of thawing in relation to actual insemination, temperature changes from thawing to insemination, site and speed of semen deposition, and sanitary procedures. If semen is purchased from a reputable supplier, it is unusual for the cause of the infertility to be poor-quality semen, although transport and storage factors should be considered.

PREGNANCY DETERMINATION IN CATTLE

Pregnancy determination is recommended to maximize breeding efficiency. In beef herds, the breeding season (natural service or AI) is ideally fixed at a length of 60–70 days. This gives the average cow two or three services to conceive. Cows that are not pregnant or were bred late should be identified; if kept in the herd, they will calve later in the season. Maintenance costs are significant, although they vary widely by farm and by year.

Pregnancy determination of beef cows should be done shortly after the breeding season is over (eg, 45–60 days); if the breeding season starts June 1 and ends early in August, it can be done during late September while the cows still have plenty of flesh from summer pasture. It is then possible to profitably market nonpregnant cows before expensive winter feeding starts. When excessive returns are noted or bull performance issues are suspected, early pregnancy determination by ultrasound 50 days after the start of the breeding season can be invaluable. A target of 65% of cows should be pregnant to the first cycle in beef herds, and so by 50 days this can be measured as pregnancy determination success 30 days after breeding. Supplementary AI or new bull power can be implemented with poor success, alongside further investigation.

Dairy cows should be examined to determine their pregnancy status and if found open can be synchronized into estrus with prostaglandin $F_{2\alpha}$, or timed AI. In cows that are open or that have not been detected in heat, the decision on which hormones to use to induce estrus is based on evaluating the ovaries for CL, follicles, and ovarian cysts. The most common method to determine pregnancy and evaluate ovaries is transrectal palpation and, increasingly, the use of ultrasonography. Ultrasonography may have the following advantages: 1) Nonpregnant cows can be found earlier (28–32 days after breeding), and diagnostic information on the status of the ovaries and uterus can be obtained. 2) The viability of the embryo or fetus can be assessed, eg, by visualization of a fetal heart beat. 3) Twins are more readily detected. 4) Sex of the fetus can be determined. 5) Age of the conceptus can be estimated more accurately. 6) The producer can be shown the conceptus, which could be reassuring in herd cases of embryonic losses.

Other herd health examinations should be considered while the cows are being checked for pregnancy. These include an accurate evaluation of body condition, the reproductive tract, teats and udder, feet and legs, teeth, and early neoplastic eye lesions. Vaccinations, internal and external parasite control, and processing of beef calves also can be done at this time.

Embryonic Death, Abortion, and Abnormal Fetal Development in Cattle

Pregnancy may be terminated prematurely, resulting in abortion due to death of the conceptus or failure of the uterine environment to support the fetus. Abnormal fetal development may result in abortion or in a calf that dies soon after birth. Many cases of bovine abortion are not diagnosed. (Also see Abortion in Large Animals.)

Etiology

Viruses, bacteria (including rickettsia and chlamydiae), molds, protozoa, or other infectious agents may attack the placenta or the fetus, or both. Some of these microorganisms reach the uterus hematogenously; others (such as venereal infections) are contracted during mating.

Infectious abortion may be sporadic or a herd problem. Herd problems usually are associated with significant losses and may be caused by infectious bovine rhinotracheitis, bovine viral diarrhea, brucellosis, leptospirosis (various serotypes), campylobacteriosis, trichomoniasis, anaplasmosis, ureaplasmas, mycoplasmas, and others not yet identified.

Mycotic abortion usually is caused by *Aspergillus* or *Mucor* spp, which reach the uterus hematogenously and cause abortion in late gestation. In many of these fetuses, the skin is not affected; in others, ringworm-like lesions are seen. The placenta frequently is severely affected with necrosis of the cotyledons and thickening of the intercotyledonary areas. Diagnosis is based on identification of the fungus through culture of the fetal or placental tissues, histologic examination of these tissues, or direct examination of cotyledons after clearing with potassium hydroxide solution. These abortions are almost always sporadic, and the only means of control is to reduce exposure to the fungi.

Sporadic losses may result from *Listeria* sp (a bacterium occasionally present in silage when pH is >7); miscellaneous bacteria such as *Haemophilus* sp, *Trueperella pyogenes*, *Staphylococcus aureus*, *Bacillus cereus*, *Pasteurella multocida*, *Pseudomonas aeruginosa*, *Streptococcus bovis*, *Chlamydia* sp, and others; or viruses (eg, bluetongue).

Noninfectious causes of abortion are numerous; the most common include 1) recessive or lethal genes (or both) such as cervical vertebral malformation, hydrocephalus, osteopetrosis ("marble bone" disease), arthrogryposis ("crooked calf" syndrome), and several others, some not fully identified; 2) toxins (eg, excessive nitrates from feed or water), certain pine needles, poisonous plants (eg, lupine, locoweed), or mycotoxins (moldy feeds); 3) hormonal imbalances in the pregnant dam; 4) injuries affecting the pregnant cow; and 5) nutritional deficiencies, particularly of vitamin A, vitamin E or selenium (or both), iodine, and manganese.

Heat stress in cattle can cause early embryonic death and lower the herd pregnancy rate. The mechanism by which heat stress affects embryonic survival is complex. Heat stress can disrupt early embryonic development. Effects of heat stress on embryonic survival decrease as embryos advance in development. Heat stress at day 1 or days 1–3 after breeding reduces embryonic survival. In contrast, heat stress of superovulated cows at day 3, 5, or 7 after estrus did not affect embryonic development in one study.

Diagnosis

Accurate diagnoses of reproductive loss contribute to the cumulative herd history and provide criteria to evaluate the impact on herd performance and the need for implementation of preventive measures. Laboratory assistance is needed in most cases. Carefully selected, properly preserved, quality specimens should be submitted to a diagnostic laboratory for analysis. Even with these, the exact cause of an abortion may not be detected, especially if it is noninfectious. Laboratory diagnosis of abortion may include serology and examination of the fetus and placenta. However, a definitive diagnosis of bovine abortion remains challenging because, in many cases, the causative agent may have challenged a cow months previously and may no longer be present when abortion occurs.

Defective newborn calves can be recognized only by a thorough examination and sometimes only after some time has passed.

Prevention and Control

Several factors are critical to prevent and control abortion and development of defective calves. Measurement and management of herd health status and planning to control infectious disease risk with appropriate boundary and purchased stock biosecurity alongside vaccination programs are essential. A balanced nutritional program helps control losses associated with mineral or vitamin deficiencies and poor-quality feeds, including moldy grains and forages. Genetic selection and accurate record-keeping help to detect and eliminate bloodlines that prove to be carriers of recessive or lethal genes. Appropriate housing and handling facilities decrease the incidence of accidents and provide an environment conducive to health. The cattle producer and veterinarian should work together to assess the herd's reproductive performance, tailor a vaccination program to the herd's specific needs, and diagnose and control potential herd problems.

For successful abatement of heat stress, the environment of the cow must be modified to maintain the cow within a normal temperature range of 101.3°–102.8°F (38.5°–39.3°C). Common approaches include providing shade (to intercept solar radiation) and fans or sprinklers to promote evaporative cooling. Tactically avoiding breeding at hottest times of the year may be advisable.

Infectious diseases in a herd can disrupt and reduce reproductive efficiency by causing embryonic or fetal death, abortion, or illness and death of neonates. A complete vaccination program will not eliminate reproductive problems but may prevent or reduce losses associated with specific infections (see Table: Vaccination/Immunization Program to Protect Against Prepartum Diseases of the Breeding Herd andVaccination/Immunization Program to Protect Against Diseases of the Neonatal Calf).

Vaccination/Immunization Program to Protect Against Prepartum Diseases of the Breeding Herd

	Disease	**When to Vaccinate/Immunize**
	Disease	**When to Vaccinate/Immunize**
Heifers	Brucellosis	Calfhood
	IBR[a], BVD[b]	Before weaning and before breeding
	Campylobacteriosis, leptospirosis, trichomoniasis	Before breeding
Cows	IBR, BVD	May booster early before breeding
	Campylobacteriosis, leptospirosis, trichomoniasis	Each year, before breeding
Bulls	IBR, BVD	Calfhood and booster before first breeding
	Campylobacteriosis, leptospirosis	Each year, before breeding

Vaccination/Immunization Program to Protect Against Diseases of the Neonatal Calf

	Disease	**When to Vaccinate/ Immunize**
	Disease	**When to Vaccinate/ Immunize**

Heifers and cows	Rotaviruses and coronaviruses, *Escherichia coli* bacterins, clostridial bacterins	As on label
Calves	Rotaviruses and coronaviruses (if indicated)	As on label

The science of animal breeding is defined as the application of the principles of genetics and biometry to improve the efficiency of production in farm animals. These principles were applied to change animal populations thousands of years before the sciences of genetics and biometry were formally established. Animal breeding, controlled propagation of domestic animals in order to improve desirable qualities. Humanity has been modifying domesticated animals to better suit human needs for centuries. Selective breeding involves using knowledge from several branches of science. These include genetics, statistics, reproductive physiology, computer science, and molecular genetics.

METHODS OF ANIMAL BREEDING

The following animal breeding methods are followed for improvement of animals in dairy and poultry namely:

- Inbreeding,
- Outbreeding and
- Mutations.

Inbreeding

The crossing of closely related animals is called Inbreeding. If this inbreeding is repeated continuously, it is called Upgrading. Inbreeding is used to retain desirable genetic traits in animals. Inbreeding is required in order to retain as many traits as possible by keeping the combination of genes intact. However, the inbreeding may result in homozygous recessive genes coming together to express some harmful phenotypic traits. Many breeding scientists have observed that hybrid vigor and fertility were lost due to repeated inbreeding. Such recessive and harmful genes are removed by some special techniques without sacrificing the major quality of the animal. If the race is relatively free of such harmful recessive genes, the process of inbreeding is a safe method for improvement of animals.

Outbreeding

The crossing of distantly related animals is called Outbreeding. One of the problems the animal breeder faces in outbreeding is introduction of new genes into population. In this method it is possible to **breed** a desirable type of animal with a less desirable type and then to increase the degree of desirable traits. New and high yielding genes can be introduced into the population through outbreeding. In many cases these genes may come from a variety of stock.

Out breeding in animals is useful for different purposes:

i. To produce some valuable traits

Beef cattle may be crossed with dairy cattle to produce calves for superior veal (flesh) production.

ii. To create new breeds

A new breed is produced with desired characters from the two original breeds. This process of producing new breeds takes time. The present day breeds of animals have been developed through hybridization.

iii. To produce a hybrid of superior Vigour and Value

A Mule is produced by crossing Equus equus (mare or female horse) and Equus homonius (jack or male ass). Mules are superior to horses in strength, endurance, resistance to disease and ability to work under unfavorable conditions. When a female mule is crossed with a jack, a colt is produced.

Mutations

The new traits into populations can also be induced through mutations. Since most of the mutations are harmful and the process of induction of mutations is quite expensive, this method of improvement of animals is impractical. It has been reported that a sheep in New England mutated in the direction of having shorter legs (a desirable quality) and formed the basis for racial improvement of sheep. Representative examples of improvements in animals by mutations include poultry that is resistant to white diarrhoea, increased egg production in fowls, increased fat content in milk and better meat yielding in Turkeys etc.

BREEDING OF CATTLE AND BUFFALOES

Reproduction is an important consideration in the economics of cattle production. In the absence of regular breeding and calving at the appropriate time, cattle rearing will not be profitable. A healthy calf each year is the usual goal. This is possible only by increasing the reproductive efficiency of the animals.

Successful reproduction encompasses the ability to mate, the capacity to conceive and to nourish the **embryo** and deliver the viable young ones at the end of a normal gestation period. In fact, interruption in this chain of events leads to failure of the cow either to conceive or the embryo to die or to have a premature delivery of the fetus.

The reproductive efficiency is a complex phenomenon controlled by both genetic and non-genetic factors, the non- genetic factors being climate, nutrition, and level of management. The reproductive efficiency varies not only between species and breeds but also among the animals within the same breed. Even the best feeding and management cannot coax performance beyond the genetic limit of an inferior animal. Improving the genetic merits of livestock populations is important at all levels of management. A sound breeding programme is a necessary part of the total animal production system.

Breeding Efficiency

The factors which influence the breeding efficiency of cattle are as follows:

Number of ova

The first limitation on the breeding efficiency of fertility of an animal is the number of functional ova released during each cycle of ovulation. Ovulation is the process of shedding of ovum from the Graffian follicle. In the case of cow, usually a single ovum is capable of undergoing fertilization only for a period of 5-10 hours. Therefore, the time of mating insemination in relation to ovulation is important for effective fertilization.

Percentage of fertilization

The second limitation is fertilization of ova. Failure to be fertilized may result from several causes. The spermatozoa may be few or low in vitality. The service may be either too early or too late, so that the sperms and eggs do not meet at the right moment, to result in fertilization.

Embryonic death

From the time of fertilization till birth, embryonic mortality may occur due to a variety of reasons. Hormone deficiency or imbalance may cause failure of implantation of fertilized ova which die subsequently. Death may occur as a result of lethal genes for which the embryos are homozygous. Other causes may be accidents in development, over-crowding in the uterus, insufficient nutrition or infections in tile uterus.

Age of first pregnancy

Breeding efficiency may be lowered seriously by increasing the age of first breeding. Females bred at a lower age are likely to appear stunted during the first lactation, but their mature size is affected little by their having been bred early.

Frequency of pregnancy

The breeding efficiency can be greatly enhanced by lowering the interval between successive pregnancies. The wise general policy is to breed for the first time at an early age and to rebreed at almost the earliest opportunity after each pregnancy. In this way the lifetime efficiency is increased. Cows can be rebred in 9-12 weeks after parturition.

Longevity

The length of life of the parent is an important part of breeding efficiency, because the return over feed cost is greater in increased length of life. Also, it affects the possibility of improving the breed. The longer the life of the parents, the smaller the percentage of cows needed for replacement every year.

Management Practices to Improve Breeding Efficiency

Some of the management suggestions which will tend to improve breeding efficiency of cattle are listed are:

- Keep accurate breeding records of dates of heat, service and parturition. Use records in predicting the dates of heat and observe the females carefully for heat.
- Breed cows during near the end of mid heat or heat period.
- Have females with abnormal discharges examined and treated by veterinarian.
- Call a veterinarian to examine females not settled after three services.
- Get the females checked for pregnancy at 45 days to 60 days after breeding.
- Buy replacements only from healthy herds and test them before putting them in your herd.
- Have the females give birth in isolation, preferably in a parturition room and clean up and sterilize the area once parturition is over.
- Follow a programme of disease prevention, test and vaccination for diseases affecting reproduction and vaccinate the animals against such diseases.
- Practice a general sanitation programme.
- Supply adequate nutrition.
- Employ the correct technique.
- Provide suitable shelter management.
- Detect silent or weak heat, by using a teaser bull.

Selection and Culling

Selection and culling are the two sides of the same coin. Selection is the process in which certain individuals in a population are included for becoming the parents of the next generation. Automatically some are excluded for the purpose which are culled. Natural selection has been going on since ages where animals which were stronger, which had better survivability and which were in more unison with the environment around them, found a better chance to reproduce.

Thus certain genes for certain characters got more chance to be selected to form individuals in the subsequent generations. Since domestication of cattle, man has been looking for superior phenotypes in traits useful to him and selecting such animals to form the parental generation. This is manmade artificial selection. Now man has progressed one step further in making estimates of genotypes from the study of phenotypes and making use of that information (in artificial) selection.

A. Selection Methods

There is only one way to select and that is to "keep the best and cull the poorest. The various selection methods are techniques for identifying or estimating the genetic values of individual candidates for selection. The procedure discussed here apply to selection for quantitative trails.

Performance Testing

Performance test is a measure of the phenotypic value of the individual candidates for selection. Since the phenotypic value is determined by both genetic and environmental influences, the performance test is an estimate, not a measure of the genetic value. The occurrence of this estimate depends upon the heritability of the trait i.e. on the degree to which the genetic value is modified by the environmental influences.

Advantages

1. Among simple procedures, the performance test is the most accurate.
2. Environmental influences can be minimized by testing candidates for selection in the same pen or in similar environmental conditions.
3. The measure is direct, not on a relative basis.
4. All candidates for selection can be tested in contrast to progeny testing where only a parent can be tested.
5. Generation intervals are usually short.
6. Testing can usually be done on the farm under normal management conditions.

Disadvantages

- Accuracy become low when heritability is low.

- Phenotypes are not available for one sex or in sex limited traits such as milk yield.
- Traits which are not expressed until maturity may become expensive or difficult to manage by performance tests since most selection decisions must be made before maturity.

Performance tests should be the backbone of most selection programmes. Although much publicity has been given to other selection methods, it remains a fact that most of the progress in livestock improvement to date has been due to selection on the individual's own phenotype i.e. performance test.

B. Pedigree Selection

A pedigree is a record of an individual's ancestors including its parents. This information is valuable because each individual possesses a sample half of the genes from each parent. If we can precisely know an individual's phenotype, little is gained by considering pedigree in selection. Pedigree considerations are useful when we do not have sufficient accurate records of production of the individual. Also, it is useful in the early selection when the traits in question might not have expressed themselves. It is also useful for selection of males when the traits selected for are expressed only by the female such as milk production in dairy cattle.

Advantages

- It provides information when performance tests are not available for the candidates.
- It provides information to supplement performance test information.
- It allows selection to be completed at a young age. Pedigree records may be used to select animals for performance or progeny testing in multi-stage selection scheme.
- It allows selection of bulls can be selected on the milk records of their female relatives.

Disadvantages

- Accuracy, relative to alternative selection procedures is usually low.
- Too much emphasis on relatives, especially remote relatives, greatly reduces genetic progress.
- Progeny of favored parents are often environmentally favored.
- Relatives often make records under quite different environments, thus introducing nonrandom bases into the selection system.

C. Progeny Testing

In this method we evaluate the breeding value by a study of the expression of the trait in its off springs. Individuality tells us what an animal seems to be, his pedigree tells us what he ought to be, but the performance of his progeny tells us what he is.

Progeny testing is, of course, a two-stage selection system because some preliminary selection determines which animals first produce progeny followed by further culling of these which produce poor progeny.

Advantages of progeny testing

- High accuracy when many progeny are obtained.

Disadvantages progeny testing

- Long generation interval.
- Requires high reproductive rate.
- Low selection intensity.

D. Show Ring Selection

Selection on the basis of show ring performance has had considerable value in the past. Essentially this selection has been directed towards bringing the conformation of the animal to some ideal conformation.

This improvement has been based on two goals:

- improvement conformation, and
- correlated response.

Improvement of conformation has economic value because a part of the sale price is determined by the conformation of the individual. The ideal type was chosen so that, in the opinion of the judges, the animal possessing this conformation was most likely to be a profitable producer. In other words, the judges were attempting to stress traits of conformation which are corrected with productive ability.

With the advent of record keeping it was found that direct selection for performance traits resulted in much faster progress than selection through correlated conformation traits. Also, when subjected to intensive study, many of the correlations between performance and show ring were found to be of non-genetic origin.

If the correlations are of genetic origin, direct selection for performance should improve conformation as well as the reverse situation. The show ring has been a good forum for discussion of what constitutes ideal type and good management and has produced dramatic changes in the conformation of some species.

This has resulted primarily from education of the breeders, however, for most animals which are presented in the ring are good and selection differential among these animals is usually so small as to produce little change.

Disadvantages of show ring selection

- Emphasis is usually placed on traits of little economic importance.
- Clever fitting and showmanship can mask defects of various kinds.
- Differences between exhibited animals are usually small.
- Conformation and production traits usually have low genetic correlations.

Choosing Traits for Selection

Many factors enter into the choice of traits to be selected for. The following ones are the most important.

- The goal of the selection programme
- The habitability of the traits
- The economic value of improvement in each trait.
- The range in variation of each trait.
- Correlation among the traits.
- The cost of the selection programme.

a. Selection Goals

Often the goal of the selection programme makes the choice of traits quite obvious. Similarly, the breeder of dairy cattle generally sets out to breed cows with superior milk production characteristics. Thus, his choice of traits is specified by his selection goals.

b. Heritability

Heritability is defined to be the fraction of the superiority of parents which is, on the average, transmitted to their off-springs. To explain habitability in simpler words: Heritability tells us how much of the observable differences in the animal is caused by genes and how much by environment.

Heritability for the same characteristics may vary from one population to another and also may vary from one characteristic to another even ink the same population. The ability to recognize the breeding values or transmitting abilities of animals is closely associated with heritability. If the heritability is high for a trait, we can proceed straight way to adopt a system of mass selection of superior animals, with little attention to pedigree information, collateral relatives, progeny test or inbreeding and genetic improvement in that trait is low, genetic progress may be disappointing with mass selection and greater attention should be paid to pedigree records, family information and use of progeny tests.

c. Variability of the trait

Selection operates on the variability in expression of the trait uniform for a trait, there will be little selection response because any selected groups of parents will not be much better than those not selected. Some traits are much more valuable than others, thus the innate variation of the traits should be carefully considered in choosing traits for selection.

Variation can be increased by improving exotic types and sometimes this can result in new combination of genes which are superior to either parent type.

d. Correlated traits

Sometimes traits tend to be inherited together. These correlations may arise in several ways.

- The traits may be of different measures of some underlying trait. For example. Weight and height are both measures of body size, thus taller animals are usually heavier and these two traits are said to be correlated.
- If the same genes produce response in several traits, those traits will be correlated. This condition is referred to as pleiotrophy.
- Correlated responses are common. Selection for increased milk yield produces a correlated decrease in the percent of fat in the milk of dairy cows. Thus, both direct and correlated responses result from selection and some correlated responses are positive while others negative.
- Correlated response may be advantageously used in selection programme. For example feed efficiency is expensive to measure because it requires both weight gain and feed intake on each individual, whereas weight gain requires neither feed weight nor individual feeding.

Systems of Breeding

The ultimate aim of the breeder is to evolve outstanding and improved type of animals which can render better service to man. Selection and system of breeding constitute the only tools available to the breeder for improvement of animals. Since new genes cannot be created though they can be recombined into more desirable groupings.

Systems of breeding has been broadly divided as under:

- In breeding -breeding of the related animals.
- Out breeding -breeding of the unrelated animals.

Inbreeding

Inbreeding is a mating system in which individuals mated are more closely related than the average of the population from which they come. It means the mating of males and females which are related. Animals deemed to be related only when they have one or more ancestors in common on the first 4-6 generations of their pedigree. The intensity of inbreeding depends upon the degree of relationship. Close inbreeding denotes mating of closely related individuals like dam to son (mother x son) or sire to daughter (father x daughter) or full brothers to full sisters.

In breeding makes more pairs of genes in the population homozygous. Wherever there is inbreeding, there will be one or more common ancestors from which, part of the gene samples (gametes) have arisen.

Estrus Period

- Pro estrus: 2 or 3 days
- Estrus: 12 to 18 hours
- Ovulation: 12 to 16 hours after the end of estrus
- Estrous cycle length: 21± 3 days

Puberty

- Puberty is the stage at which animal becomes sexually mature and secondary sex characteristics become conspicuous.
- The term sexual maturity means that the animal is capable of reproduction.
- Puberty is the age at which the first estrus occurs in the heifer and the bull starts giving semen with viable sperms.
- The reproductive organs undergo marked increase in size at the time of puberty.
- Under good feeding a calf attains puberty approximately at 66% of adult body size.

Signs of Estrus

- Cow in estrus will be the first cow to rise in the morning.
- The cow become restless does not eat and frequently bellows and seldom ruminates.
- Sudden drop in milk production.
- Searching for male.
- Traits of homosexuality is shown in which the cow will attempt to mount other cows while other females not in estrus tend to mount the estrus cow which she permits.
- The cow is receptive to the act of mating and will stand when the bull mounts her.
- The behavior of standing quietly while being mounted by the bull or other cow is referred to as the 'standing heat' which is the surest sign of estrus.
- This extends for 14-16 hours and shows other symptom like bellowing, nervousness, anorexia, reduction in milk yield.

- Mucous discharge may be found sticking to the tail.
- In early heat the discharge is watery and copious in mid heat (standing heat) it becomes thick and sticky and in late heat it will be scanty and discolored.

Bulling

The best indicator of estrus is when any cow or heifer repeatedly stands and accepts mounting by one of her herd mates. Unfortunately, they do not do this on demand. Those responsible for estrus detection must watch for this behavior and combining what they see with their own previous knowledge/experience, to decide whether to inseminate or not.

Heat detection in buffaloes:

- Cows do mount over other cows when they are likely to come in heat and stand for mounting when they are in good heat. This is not seen in buffaloes. Buffaloes neither mount on other buffaloes nor is other buffaloes mount on buffaloes in heat. In buffaloes copious ropy hanging discharge not seen on the contrary it gets suddenly dropped and is not noticed by the owner and the discharge is scanty. Some buffaloes do not bellow and show silent heat, especially high yielding buffaloes.
- The main heat symptoms of buffaloes are as follows. The vulva becomes edematous, swollen. The lower portion of vulva looks oily. The gap is seen between vulvar lips and slight opening is seen. The wrinkles which are present in anestrus buffalo become shallow or vanish.
- The mucous membrane of vulva becomes reddish, moist and glossy.
- Mucus discharge which is not seen normally can be seen before or after estrus spontaneously.
- The color, consistency and fern pattern of mucus help in determination of correct estrus.
- Engorgement of teats in lactating buffaloes which is due to holding of milk following increased estrogen level in blood is seen when they are in heat.
- Frequent urination. The urine coming in spurting action wetting the part of skin below vulva and above udder (perineum). The drying of the urine leaves white mark on skin.
- Buffaloes in heat remain restless, off feed, raising head in a typical fashion.
- Local non-descript buffaloes bellow, become restless and remain off feed. Milk yield is reduced. The bellow is sharp and for longer duration.
- The buffaloes expose their teeth while bellowing which is very characteristic.
- The mucus discharge, in buffaloes is seen in about 49% cases. It is thin on the day of heat, become thick as the time passes and changes the color from clear to white.
- 60-70 % of the buffalo come in heat from 6 pm to 6 am (after sunset and before sunrise) and this should be borne in mind and attendant should watch the buffaloes in the evening and early morning for expression of heat symptom.

- Teaser bull (Vasectomized bull), can be used for parading in buffalo barn for detection of heat

Other methods to Detect Estrus

- Crystallization pattern of cervical mucous will show long crystals in a typical fern-like pattern
- There are many estrus-detecting devices available. They are usually attached to the tail or rump of the cow.
- Mounting causes these devices to discharge a colored fluid which can be observed afterwards even from the distance.
- 'Chin ball mating device' can be used for heat detection. It works on the same principles of a ball point pen and is fixed by means of a halter below the chain of the teaser bull. When the checking animal mount the cow in heat, the dye exuded round a spring-loaded ball of the device marking the back of the cow.
- Russian workers have developed an instrument basically consisting of an ohm meter and electrodes. When applied to the mucous membrane of the vagina, the resistance indicated on the ohm meter shows whether the cow is in heat.
- Pedometer is an instrument used to monitor the movement of animal. The principle is the activity and movement of the cow increases on the day of heat and this can be detected by means of a pedometer tide to the leg of the cow.
- The vaginal temperature can be recorded, which gives an indication about the heat. Generally during estrus, the vaginal temperature increased by about 1°Con the day of heat. Both methods are not very practicable.
- The methods described above had little applicability in developing countries due to technological and economical and manage mental reasons, close observation of signs of heat, standing heat remains the most practicable method of heat detection.
- In large farm this can be supplemented with a bull-parade using a teaser bull. A teaser may be a vasectomized bull or bull in which penis has been amputated and the urethra exteriorized.
- An intact bull can also be used by hanging a thick cloth or gunny bag curtain in front of the penis preventing entry of penis and mating.

Time of Insemination

- Ovulation takes place about 12 hours after the end of estrus. It takes another six hours for it to travel half-way down the oviduct.
- The sperm, even though reach the oviduct within minutes after insemination, must be exposed to the female reproductive tract for about 6 hours to attain the capacity to fertilize.
- This process of preparing the sperm for fertilization is known as capacitation.
- Sperm are viable for 24 hours in the female reproductive tract whereas the

ovum remains fertile for only about 10 hours after ovulation.

- This implies that mating or insemination between mid-estrus (middle of standing heat followed by another insemination in about 6 hours after that.
- As a routine practice, if a cow is seen showing signs of early heat in the morning, it may be inseminated in the evening.
- If such signs are first manifested in the evening, the cow may be bred next day morning.
- A cow is expected to show estrus in 30-40 days after calving. Cows that fail to show heat even after 50 days have generally some problem and need examination.
- It may be due to infection or malnutrition and remedial measures may be taken accordingly.
- Insemination should be done only when buffalo is in standing heat. In buffalo to understand standing heat one should know the symptoms of heat.
- Buffaloes normally are not seen standing for mounting by herd mates but standing heat can be known from the changing color of mucus discharge which is early estrus is clear and watery but in standing heat or mid heat the color is changed to little buffy with thick consistency.
- In mid heat the oedema of vulva is intense there is little gap in vulvar lips and lower lip looks oily.
- The vulvar mucus membrane is glossy reddish or pink and wet.

Signs of Approaching Parturition

- Cow will leave the herd and seek isolation.
- Loss of appetite and distress.
- Distention of teat and udder, considerable milk appears in the udder and there may be dripping of milk.
- Relaxation of pelvic ligament one day before calving, the ligament on the sides of the tail head is loosened so that hollows appear on either side of the backbone and the tail head is raised and the quarters are dropped.
- The vulva become enlarged and flabby
- Animal will be restless and will pace about often trying to kick or scratch the flank region.
- The parturition process has three stages a. preparatory stage (uterine contraction and dilatation of cervix) b. active expulsive stage c. expulsion of fetal membrane.
- Cow will deliver the calf within 12 hours after commencement of first stage and lapse in this vaginal examination of assistance by a veterinarian is required.
- Care must be taken to observe expulsion of placenta (after birth). It should be removed immediately so as to avoid cow eating it.

Package of Practice to Improve Reproductive Efficiency

- Accurate record kept is very important in ensuring reproductive efficiency I the herd.
- Their production detail like date of estrus, date of service and calving should be maintained properly.
- This data should be used to predict the probable date of heat, such animal should be watched carefully in the morning and evening for signs of heat.
- In larger dairy farm teaser bulls can be put in use.
- Complete breeding history, past performance and difficulties of a individual cow should be maintained.
- Irregular estrus and abnormal discharge should be attended immediately.
- The cows with retained placenta should be treated promptly and when such cows are put in breeding next time, the reproductive tract should be examined thoroughly for involution and possibility of infection.
- A manager should examine a cow 24 to 36 hours after service for met estrus bleeding. If it occurs under 24 hours after service, the cows were bred too late.
- If it occurs over 36 hours after service, they were bred too early during estrus. This will help in pinpointing of failure of conception.
- Cow should be examined for pregnancy 45 to 60 days after service so that if they are non-pregnant, steps can be taken to re-breed them at the earliest opportunity.
- If the conception rate under A.I is lower than natural service, time of insemination, insemination technique and quality of semen must be checked.
- Short irregular cycles indicate cystic ovaries, short and long irregular interval point to missed head.
- Silent or quiescent heat: the behavioral manifestation of heat may be very weak or imperceptible in such case. It is very common in buffaloes. But there is a normal ovulation and if inseminated at proper time the animal can conceive.
- Cows go through the normal ovarian changes of the estrus cycle except the behavioral heat and sexual receptivity.
- It is more in summer season than other seasons and more in heifer than adult animals.
- Use of balanced feed, proper summer management, use of teaser bulls can be of use tool in detecting silent heat.
- Anestrus or absence of sexual cycle may be due to under developed genitalia or due to persistent CL. In the former case follicle fail to develop and a heifer will not come to heat at all.
- One of the major causes of under developed genitalia is malnutrition. Besides there can be genetic causes.
- The second probability is the anestrous also may be due to persistent CL, due

to certain hormonal disturbances, the C L persist beyond the life expectancy in a normal cycle, thereby preventing further cycling. A common cause for persistent CL is endometritis of the uterus.

- Sometime anestrus is often observed in the early post-partum period when the lactation is strong, probably due to the influence of lactation (due to secretion of prolactin) 'e'lactational anoustnum.

Care and Management of Young Stock

- Normally newborn animals will be taken care by its mother and required little assistance.
- In case of cattle, sheep and goat immediately after birth the mucus around the nostrils should be whipped out using dry cloth or a hand full of straw can be used for this purpose.
- Calm environment should be provided to the mother and young animals for development of bond.
- The mother should be allowed to lick the newborn; if the dam fails to lick it can be stimulated by sprinkling small quantity of salt or bran over the young one.
- Immediately after birth the naval cord should be ligated with clean sterile cotton thread 1 inch from the body and tincture iodine should be applied to the naval cord.
- Within 1 hour after birth the newborn will able to stand and it should be allowed to drink adequate quantity of colostrum (first milk) which will give immunity to the newborn.
- Young animals should be housed comfortably. Adequate care should be taken to avoid housing young stock with adult stocks.
- In winter condition adequate warmth condition should be provided.
- Adequate bedding materials like straw or hay should be provided to newborn animals.
- For giving extra heat artificial light source can be utilized.
- Proper light, ventilation and hygiene should be maintained to avoid spread of disease.

Care and Management of Pregnant Animals

- Pregnant animals should be provided with extra ration to meet the requirement of fast growing fetus as well as store energy for future lactation.
- Pregnant animals should be separated in advanced stage from other non-pregnant animals.
- They should be housed separately in place called calving pen.
- Adequate bedding materials should be provided in the pregnant animals shed.

- Floor of the pregnant animal shed should be non-slippery.
- Adequate clean fresh drinking water should also be provided in the calving pen.
- In advance stage of pregnancy laxative diet should be provided.

Maintenance of Sexual Libido of Bulls

- There are several factors which can reduce libido in bulls like young or old age, inexperience, tiring exercise, or too frequent usage, semen collection at unusual places in un favorable conditions and using unsuitable fittings, faulty feeding, obesity or run down condition, inherent defects, temporary injury or chronic defect of legs, back and penis. All such problems should be rectified as soon as noticed.
- Some bulls are sensitive to artificial vagina whereas others seem able to withstand considerable rough handling.
- The well-known reflexes of mounting the cow, projecting the penis, thrusting and ejaculation can easily be retarded or even inhibited in a bull by unnatural method of handling.
- Majority of the bulls serve well in familiar surroundings and are handled by the same attendant provided these are associated with previous satisfactory experience.
- The sexual reflexes can be inhibited by painful, uncomfortable or even distractive situation.
- In a sensitive bull, inhibition may develop quickly, even when collections are taken carefully.
- Overwork is common in young bulls allowed free access to cows and heifers.
- The number of services and not the number of cows served is the important consideration. No bull should be allowed to serve each cow more than twice in a heat period.
- A young bull may be placed with 2 or 3 cow per week and it can be put into service after 2-2 ½ years of age.
- A mature bull may ejaculate many times per week without effect on libido or semen quality.
- The bull with reduced libido should be teased by delaying the service. Bulls become bored in their surroundings, particularly if in small paddock and may lose interest.
- Presence of another bull or change in the surrounding will overcome this problem.
- Summer stress leads to low sexual libido and poor semen quality, especially in purebred exotic and crossbred bulls.
- To overcome such problems during summer, bulls should be housed in cool, well ventilated dry sheds.

- Showering or splashing cold water on bull 2 or 3 times during hot part of the day and protection against direct and reflected radiation were found to be very useful.

Feeding of Mature Bulls

- A good rule to feed mature bull is to feed daily about 1 kg hay and ½ kg concentrate per 100 kg body weight.
- Thus a 400 kg bull should get 4 kg hay and 2 kg concentrate.
- These amounts should be adjusted according to the body condition of various bulls because there is individual variation in response.
- Excess fatness in mature bull should be avoided at all costs as it reduces libido and may cause severe stress and strain on their feet and legs.
- Excess calcium in bull ration can cause problem particularly in older bulls.
- When legume roughage is fed the concentrate mixture should not contain a calcium supplement.
- Generally bulls do not lose calcium and in time excess calcium may cause vertebra and other bones to fuse together.
- Therefore bulls may need a different concentrate mixture than the milch cows.

Reproductive Management of Females

Assuming that female stock are healthy, their body condition or weight at mating is the factor which exerts the greatest influence on their ability to conceive. Farmers should thus aim to manage their animals so that target weights and conditions are met. For example, beef heifers should attain 75% of their mature weight when mated at 2 years of age. Mature cows have the best chance of conceiving when reaching 100% of their mature weight at mating. Only about 50% of females will conceive when reaching 75% of their mature weight at mating.

The chances of reaching acceptable target weights are maximized when feed resources are matched with the nutritional requirements of beef cows. Thus, calving should be arranged to occur in late winter or early spring, so that cows with suckling calves, which have a high feed demand, can utilize good quality, actively growing spring and summer veld. After weaning in autumn, dry cows will have a low feed requirement and can then utilize poorer quality winter veld grazing. Diseases, and poor control of internal and external parasites in females can suppress their ability to reproduce. Farmers are therefore advised to consult with veterinarians and other experts to plan effective vaccination, dosing and dipping programmes. It is possible to improve reproductive rates by manipulating the amount of milk produced by females, or the suckling activity of their offspring. Females which produce excessively large amounts of milk tend to reproduce poorly, especially when nutritional levels are inadequate. Farmers can select against cows which produce excessive amounts of milk, and can reduce the suckling frequency of offspring, by techniques such as temporary weaning (e.g. for 48 hours in beef cattle) or restricted suckling (e.g. suckling beef cows only once a day). The improvement in reproductive rates as result of these techniques varies considerably though, and these measures should be considered only when stock are undernourished.

REFERENCES

7. Adeyemo O and Heath E. 1980. Plasma progesterone concentration in Bos taurus and Bos indicus heifers. Theriogenology 14: 411-419.
8. Agarwal S P. Rahman S A, Laumas K R. Agarwal V K and Gautam O P. 1977. Studies on steroid hormones. Progesterone concentration in the blood serum of zebu cows. Indian Journal of Animal Sciences 47: 715-719.
9. Batra S K and Pandey R S. 1983. Relative concentration of 13,14-dihydro-15-keto prostaglandin F2 a in blood and milk of buffaloes during the estrous cycle and early pregnancy. Journal of Reproduction and Fertility 67: 191-195.
10. Berardinelli J G. Dailey R A, Butcher R L and Inskeep E K. 1979. Source of progesterone prior to puberty in beef heifers. Journal of Animal Science 49: 1276-1280.
11. Bergh A, Nikula H. Dember J E, Clayton R and Huhtaniemi I. 1985. Altered concentrations of gonadotrophin, prolactin and GnRH receptors, and endogenous steroids in the abdominal testes of adult unilaterally cryptorchid rats . Journal of Reproduction and Fertility 74: 279-286.
12. Braden T D, Gamboni F and Niswender G D. 1988. Effects of prostaglandin F2 alpha-induced luteolysis on the populations of cells in the ovine corpus luteum. Biology of Reproduction 39: 245-253.
13. Caligaris L, Astrada J J and Taleisnik S. 1972. Influence of age on the release of luteinising hormone induced by oestrogen and progesterone in immature rats. Journal of Endocrinology 55: 97.
14. Carruthers T D and Hafs H D. 1980. Suckling and four-times daily milking: Influence on ovulation, estrus and serum luteinizing hormone, glucocorticoids and prolactin in postpartum Holsteins. Journal of Animal Science 50: 919-925.
15. Carruthers T D, Kesner J S. Convey E M and Hafs H D. 1978. Hypothalamic GnRH and pituitary function in suckled and non-suckled Holstein cows. Journal of Animal Science 47 (Supplement 1): 349.
16. Foley R C and Greenstein J S. 1958. Cytological changes in the bovine corpus luteum during early pregnancy. In: F X Gassner (ed.), Reproduction and infertility. Pergamon Press, New York, USA. pp. 88-96.
17. Hunter R H F. 1980. Physiology and technology of reproduction in female domestic animals. Academic Press, London, UK. 393 pp.
18. Johnson C T. 1981. Induction of parturition in cattle. Acta Veterinaria, Scandinavica, Supplementum 77: 311-319
19. Laster D B. 1977. A pregnancy specific protein in the bovine uterus. Biology of Reproduction 16: 682.

CHAPTER 2

Cattle Nutrition

Chapter Contents

Feeding diets with lowered protein content reduces nitrogen input, improves nitrogen utilization efficiency, and reduces nitrogen losses from manure. Reducing dietary protein also benefits the producer by reducing feed cost and improving overall farm profitability. These interventions, however, have to be balanced with the risk of loss in milk production. If the true animal requirements for metabolizable protein are not met, long-term production cannot be sustained.

INTRODUCTION

Animal science deals with the study of non-human animals. More specifically, the discipline emphasizes domestic animals, or animals that are controlled and cared for to some degree by humans. Major biological issues such as physiology, anatomy, and development are investigated by animal scientists. These studies are then applied to issues ranging from animal nutrition to animal reproduction. Targeted sub-disciplines include agriculture and veterinary medicine. The main area of focus in animal science is the structure and function of domesticated animals. Therefore, scientists may investigate specific body organs and their capacities as well as cellular and chemical processes that take place within the animals. More specific areas of study can result from these investigations, such as cycles of reproduction and how the animals respond to various foods. Environmental influences and human intervention play a major role in these animals' lives, so consideration of these issues is commonplace as well. For example, factors like climate and soil richness affect plant growth, and plants are the primary source of nourishment for many domesticated animals. As such, these factors are often of great interest to individuals involved with animal science. Mankind also greatly impacts animals and domesticated animals in particular, either in a personal caretaker capacity or with more large-scale considerations like habitat-destroying industrial growth or animal-based laws and regulations. Animal nutrition focuses on the dietary needs of domesticated animals, primarily those in agriculture and food production. There are seven major classes of nutrients: carbohydrates, fats, fiber, minerals, protein, vitamin, and water.

The macronutrients (excluding fiber and water) provide structural material (amino acids from which proteins are built, and lipids from which cell membranes and some signaling molecules are built) and energy. Some of the structural material can be used to generate energy internally, and in either case it is measured in joules or calories (sometimes called "kilocalories" and on other rare occasions written with a capital C to distinguish them from little 'c' calories). Carbohydrates and proteins provide 17 kJ approximately (4 kcal) of energy per gram, while fats provide 37 kJ (9 kcal) per gram, though the net energy from either depends on such factors as absorption and digestive effort, which vary substantially from instance to instance. Vitamins, minerals, fiber, and water do not provide energy, but are required for other reasons.

IMPORTANCE OF NUTRITION AND DIGESTION

Nutrition is important for a variety of reasons. Animals need the proper nutrition for growth and maintenance, and to provide energy for work and vital functions. Maintenance is the nutrition required for an animal to maintain its current weight. Energy is the ability of the body to perform functions. Proper nutrition is also needed to maintain body temperature, produce milk, reproduce, and develop proper bone structures. Without proper nutrition, animals can develop health problems, which could result in treatment costs or even fatality. Good nutrition is essential for all of the systems of an animal to function and work together properly.

Animals that do not receive the proper nutrition are more likely to develop health and reproductive problems, and be less productive and marketable. Nutrition is important to have healthy animals, and, in turn, healthy animals help ensure profitability in agricultural operations. Producers must understand not only what nutrition an animal needs but also how to supply that needed nutrition. Producers must know what ration is appropriate for a specific animal and how to provide a balanced ration. A ration is the feed an animal receives over a 24-hour period. A variety of feedstuffs, or basic ingredients, are used in rations, and producers must choose those that best suit their needs. A ration with all the nutrients an animal needs is a balanced ration.

Digestion is the process of breaking down feeds into simple substances that can be absorbed into the bloodstream and used by the body's cells. It is important for producers to understand the digestive process because it helps in selecting the proper feed for the animal's type of digestive system. The digestive system is the system in which food is acted upon through physical (chewing) and chemical (stomach acids) means. It includes all parts from the mouth to the anus. Understanding the digestive system can help the producer become more efficient in feeding, resulting in a more cost-effective livestock enterprise. It also helps the producer have a better understanding of animal health and problems that may occur. Understanding the chemical and physical changes that take place after an animal eats is important in noticing health problems related to digestion.

Basics of Animal Physiology

Physiology refers to the systems and functions of the body. What are the systems in an animal's body? The major ones include the skeletal, muscular, circulatory, respiratory, nervous, urinary, endocrine, digestive, reproductive, and mammary systems.

Skeletal System

The skeletal system consists of bones and cartilage that give the body shape. It protects internal organs and provides support for the body. Bone is the major component of the skeletal system, which consists partially of calcium and phosphate. The material inside bone is bone marrow, which produces blood cells. Cartilage provides support and serves as a cushioning effect for joints.

Circulatory System

The main components of the circulatory system are the heart, arteries, capillaries, and veins. The circulatory system delivers food and oxygen to the cells of the body and eliminates waste products. It contains blood, which consists of red blood cells, white blood cells, and platelets. Red blood cells carry oxygen while white blood cells help the body fight disease.

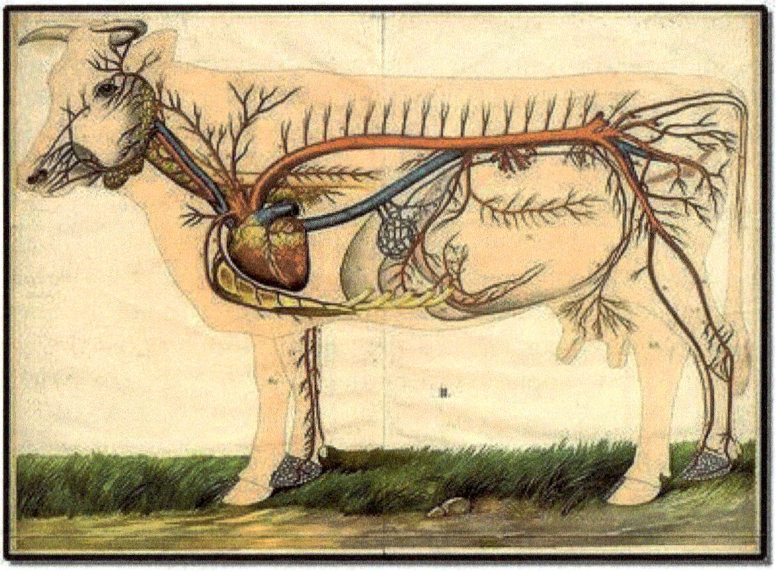

The heart, a muscular organ, pumps blood through the system. The vessels that carry blood from the heart to the body are arteries. Capillaries are small vessels that carry blood to individual cells where oxygen and nutrients are delivered, and carbon dioxide and wastes enter the blood. Veins carry blood back to the heart.

Urinary System

The urinary system, also known as the excretory system, consists of the kidneys, bladder, urethra, and ureters. The purpose of the urinary system is to remove waste products from the blood. The urinary system filters waste products and some water through the kidneys. The urine then goes through the ureters to the bladder where it is emptied through the urethra.

Muscular System

The muscular system allows movement and necessary functions such as circulation, digestion, and breathing. It is made up of muscles and connective tissue, such as tendons. The muscular system also helps provide support and shape. Muscles consist largely of protein and are the lean meat of an animal that is used as food for human consumption.

Muscles can be voluntary or involuntary. Voluntary muscles are those that can be controlled. When an animal walks or chews it is using voluntary muscles. Involuntary muscles are automatic and operate without any thought on the part of an animal. Muscles that operate the heart, intestines, and lungs are involuntary.

Respiratory System

The major parts of the respiratory system are the nostrils, nasal cavity, pharynx, larynx, trachea, and lungs. When using the respiratory system, animals breathe in needed oxygen and breathe out carbon dioxide. Air first enters the respiratory system though the nostrils or mouth. Then the air passes through the nasal cavity to the pharynx. Finally, air goes through the trachea and enters the lungs. In the lungs, air is exchanged for carbon dioxide. When animals breathe out, this carbon dioxide is removed from the body.

Nervous System

The nervous system allows the body to communicate by sending, receiving, and interpreting signals. The nervous system is composed of the brain, spinal cord, and neurons. The nervous system can be divided into two parts: the central nervous system and the peripheral nervous system. The central nervous system includes the brain and spinal cord. It is responsible for receiving information from nerves that pass through the spinal cord and others that control the senses of hearing, sight, smell, touch and taste.

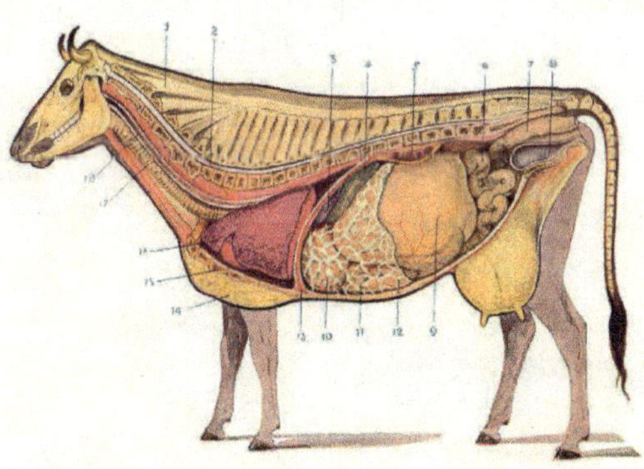

The peripheral **nervous system** connects the central nervous system to the rest of the body. The peripheral nervous system is the system that is responsible for voluntary muscle control. The autonomic nervous system is a part of the peripheral nervous system and controls automatic body activities, such as the heart muscle.

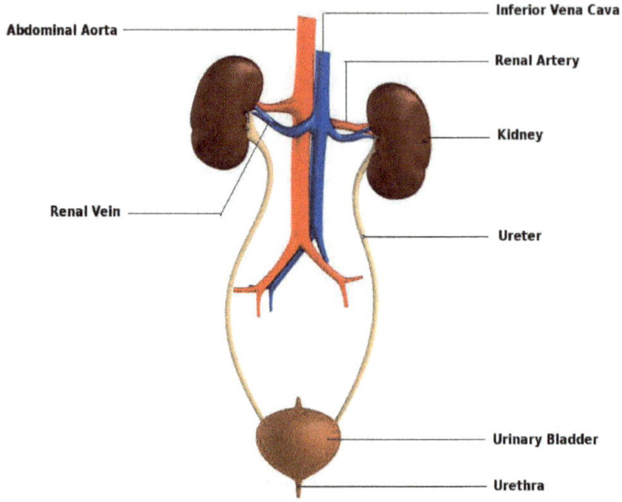

Endocrine System

The endocrine system consists of endocrine glands that release hormones into the body. Hormones are chemicals necessary for the regulation of many bodily functions. For example, hormones regulate growth and reproduction.

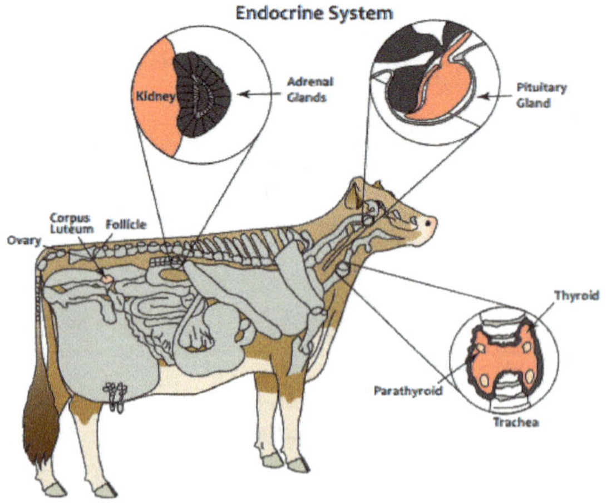

Digestive System

What happens to food after it is eaten? The digestive system breaks food down so that it can be used by the body. Digestive systems can vary depending on the species of animal. Some have very simple stomachs and others are more complex. The types of digestive systems are discussed later in this unit.

Reproductive and Mammary Systems

Have you ever seen a newborn foal or calf? The reproductive system is what enables reproduction to occur. The male and female reproductive systems differ in animals. The female reproductive system provides an environment for gestation. In female animals, the ovaries produce eggs, which are usually released on a regular cycle through ovulation. Eggs can become fertilized when mating occurs during a certain period of the ovulation cycle.

Classes and Sources of Nutrients

What nutrients do animals need? There are six major classes of nutrients: water, protein, carbohydrates, fats, minerals, and vitamins. Nutrients can be obtained from various sources. Common sources of nutrients for agricultural animals include roughages and concentrates. Roughages include feed that is high in fiber and low in energy, such as grasses, hays, and silages. Concentrates include feed that is high in energy and low in fiber, such as grains.

Water How long can one survive without water? Not long. Water is vital for all animals and makes up more than one-half of most animals' bodies! It forms the basis of fluids in the body and is essential for processes such as digestion, blood circulation, and waste elimination. Water helps dissolve nutrients and also regulates body temperature through perspiration and evaporation. Animals need a fresh, clean source of water. Most of the water an animal consumes comes from the water it drinks. Water also can be obtained from feed sources such as forage. Water is essential, especially for working animals or during hot weather. During hot weather or when working hard, horses can require 20 or more gallons of water per day.

MONOGASTRIC DIGESTIVE SYSTEM

Feeding animals correctly, in terms of both the amount and type of feed supplied, is extremely important for production, since it helps to ensure that the animal is healthy and productive. It is therefore important to understand exactly how the digestive system processes feed so the animal can use nutrients properly.

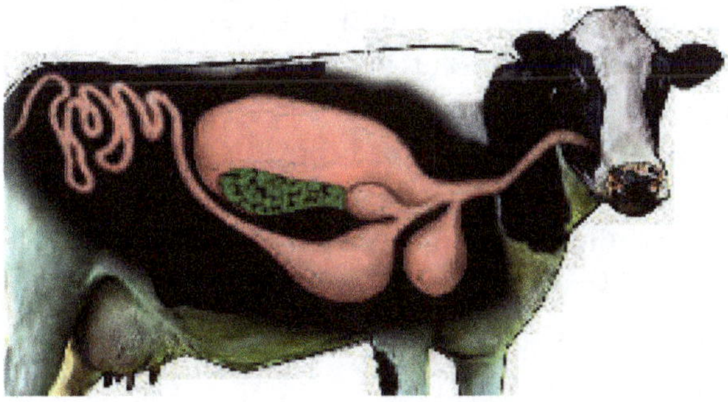

Figure: Digestive system of cow.

Monogastric Animals

The term "monogastric" refers to the structure of the stomach. In a monogastric digestive system, the stomach has a simple structure consisting of a single compartment. A number of species have a monogastric digestive system, including swine, horses, dogs, rabbits, and fowl. All of the mammals listed here have similar systems, although some minor differences do exist between them. Fowl, however, have a digestive system that differs from the others, including organs not found in the other species.

Parts of the monogastric digestive system as feed moves through the monogastric digestive system (also termed the alimentary canal) of mammals, it passes through a number of different parts (Figure)

Digestion begins when feed enters the mouth. The feed is then carried through a tube called the esophagus to the stomach. When leaving the stomach, the feed moves into the first section of the small intestine, which is called the duodenum. After traveling through the rest of the small intestine, what remains of the feed is first emptied into the part of the large intestine named the cecum and then into the colon. Finally, the waste products are passed into the lower end of the large intestine, which is referred to as the rectum, and out of the body through the anus. While they are not a part of the digestive system, the pancreas, gall bladder, and liver also play a role in digestion. They release substances that aid the digestive system in performing its function. However the horse, unlike the other animals, does not have a gall bladder. The path followed by feed in fowl differs somewhat due to the different parts of its digestive system (Figure). Feed enters the system through the beak and passes through the esophagus, which is called the gullet, to an organ referred to as the crop. From the crop it passes into the glandular stomach, or proventriculus, and then the gizzard, or ventriculus, before entering the small intestine. At the point where the small intestine and large intestine join are two ceca. After passing the ceca, the undigested material moves through the large intestine to the cloaca and exits the body through the vent.

Functions of the Parts

The parts of the mammalian monogastric digestive system can be divided into four sections based on their functions. The sections consist of the parts leading to the small intestine, the small intestine, the large intestine, and the parts allowing the exit of solid wastes, or feces. The parts leading to the small intestine reduce the size of feed particles. The process begins in the mouth as the teeth are used to grind feed into smaller pieces that are moved to the throat by the tongue. The tubelike esophagus carries these pieces to the stomach, a muscular organ that stores food before it passes into the small intestine. Digestion, which began in the mouth, continues in the stomach. The primary function of the small intestine is to further break down and absorb the nutrients found in food. The digestive processes continue in the duodenum. Nutrients are absorbed into the bloodstream through the intestinal walls in the remainder of the small intestine. The next section of the digestive system, which consists of most of the large intestine, is designed to absorb water and mold indigestible feed wastes into a solid form. The cecum, also referred to as the blind gut, contains microorganisms. Its function varies depending on the species. In some species it has little or no effect on digestion. In the horse and rabbit, however, the cecum is enlarged, which allows the microorganisms to ferment roughage, which is feed (like hay, grass, and grass products) that has a lot of fiber and few digestible nutrients. The fermentation breaks down the roughage for absorption into the blood. These two species are therefore able to make efficient use of plant materials in their diets. After passing through the cecum, the feed enters the colon, which absorbs water and forms the feces. Some nutrients are absorbed in the colon as well. The parts allowing the exit of the feces are the rectum and the anus. The rectum carries the feces to the anus. The anus passes the solid wastes out of the body. The digestive system of fowl functions somewhat differently. Fowl have no teeth with which to chew and instead use the beak to break feed down for swallowing. After passing through the gullet, feed is stored in the crop and softened for digestion before it enters the stomach. Digestion begins in the glandular stomach. The feed then passes into the gizzard where it is ground by muscular contractions before entering the small intestine for digestion and absorption. The two ceca contain **bacteria** but have little role in digestion; fowl therefore cannot digest fibrous feeds like hay. The large intestine carries wastes to the cloaca, the junction of the reproductive and digestive systems.

Digestive juices and enzymes play a major role in digestion. Digestive juices are fluids that are secreted into the digestive system from glands or tissues. Substances call enzymes are found in these fluids. An enzyme speeds up chemical reactions, allowing feed to be digested and absorbed. Table 3.1 shows digestive juices and **enzymes** found in the monogastric digestive system. In mammals, digestive juices begin to break down feed as soon as it enters the mouth. Saliva from the salivary glands in the mouth lubricates the feed for passage into the digestive system.

Table 3.1 - Digestive Juices and Enzymes

Type of Digestive Juices	Enzyme
Saliva	Amylase
Gastric	Pepsin Rennin Lipase
Pancreatic	Amylase Lipase Trypsin Chymotrypsin
Bile	No enzyme present
Intestinal	Peptidase Maltase Sucrase Lactase

It also contains the enzyme amylase. As saliva mixes with the food, amylase begins the breakdown of starches into sugars that can be absorbed by the body. It continues to act until neutralized in the stomach. Once the food moves into the stomach, other digestive juices begin to work on the feed. One of the chief components of the gastric juices secreted by the stomach wall is diluted hydrochloric acid (HCl), a corrosive substance that stops the action of amylase. Another component of the juices is gastrin, which stimulates the stomach to produce more gastric juices. The gastric juices also include the enzymes pepsin, rennin, and gastric lipase. Pepsin begins the digestion of proteins, rennin curdles a protein in milk, and lipase begins the breakdown of fats. Mucus is also present in the stomach to protect the stomach lining from the acidic environment that is created by the gastric juices. Other digestive juices are released into the small intestine.

The pancreatic juices contain several enzymes, including pancreatic lipase, which is aided by bile in the further breakdown of fats; trypsin, which breaks down proteins; chymotrypsin, which also acts on proteins; and amylase, which continues to work on starch. In addition, intestinal juices containing the enzymes peptidase, maltase, sucrase, and lactase are produced in the small intestine itself. Peptidase breaks down proteins, while maltase, sucrase, and lactase work on sugars and starches. In fowl, digestive juices also convert the nutrients in feed into a usable form. Little saliva is present in the mouth, but saliva and secretions from the wall of the crop soften the feed for digestion. In the proventriculus, gastric juices containing HCl and pepsin are secreted, and the pepsin begins the digestion of proteins. The digestive juices are further mixed with the feed as it is ground in the gizzard. In the small intestine, bile and pancreatic juices with amylase, trypsin, and lipase are released; they function in the same way as they do in mammals. The intestinal juices, which contain peptidase, maltase, sucrase, and lactase, also have the same functions as in the mammalian system.

RUMINANT DIGESTIVE SYSTEM

In contrast to an animal with a monogastric digestive system, a ruminant has four stomach compartments and can utilize some feeds more efficiently than a monogastric animal. Ruminants are important to the animal industry because they can use hay and pasture productively.

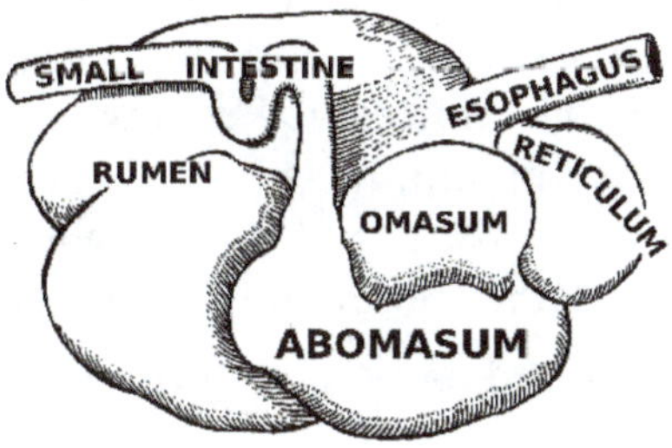

Parts of the Ruminant Digestive System

Two common livestock species have a ruminant digestive system. Cattle, both dairy and beef, and sheep are ruminants. The digestive parts and their functions are similar in both species. Some of the parts of the ruminant digestive system are the same as those of the monogastric digestive system, but other parts are very different. As in the monogastric digestive system, food enters into the system through the mouth and then passes through the esophagus to the stomach. In the ruminant, the stomach is divided into four compartments, called the rumen, reticulum, omasum, and abomasum. Depending on the type of feed, it may pass through all or some of these compartments as it moves through the stomach. When the stomach completes its functions, the feed moves to the small intestine and proceeds through the cecum, colon, and rectum of the large intestine to the anus. The digestive system of cattle is illustrated in Figure.3.2

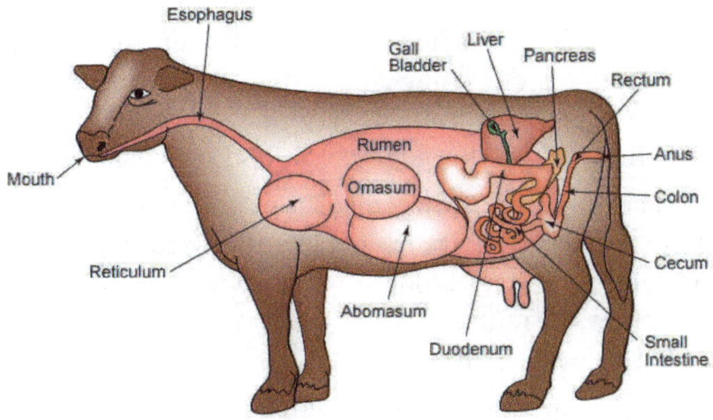

Figure - Digestive system of cattle.

Functions of the Parts

Each of the parts of the ruminant digestive system aids in the process of extracting the nutrients needed by the animal from feed. Although the parts of the system have some similarities in function to those of the monogastric digestive system, the process of digestion is much more complex due to structure of the stomach.

Mouth – The mouth carries feed into the digestive system. In cattle, the tongue grabs the grass or other feed and brings it into the mouth. The tongue also moves feed to the throat. The cow has front teeth only in the lower jaw, with a dental pad in the upper jaw; when grass is pulled into the mouth, the teeth cut the grass against the dental pad. The mouth also has upper and lower back teeth for chewing cud and other types of feed. Sheep use their lips to bring food into the mouth when grazing.

Esophagus - Feed is transported back and forth from the mouth to the stomach through the esophagus.

Stomach – Like the monogastric stomach, the main function of the stomach of the ruminant is to break down feed. Each of the four stomach compartments has its own unique function. Figure illustrates the four compartments of the stomach.

Rumen - The rumen, which is also called the paunch, is the largest of the stomach compartments. It helps to break feed down so that it may be digested.

Reticulum – The reticulum is also called the honeycomb because of the texture of the inner wall of the compartment. It pumps roughage back to the mouth through the esophagus for rumination, which is the racing of feed in the form of the cud. The reticulum also works with the rumen in the breakdown of feed.

Omasum – The function of the omasum, or manyplies, is not fully understood. However, scientists have found that the omasum absorbs some water and is involved in the absorption of nutrients.

Abomasum – The abomasum is referred to as the true stomach. Its functions are very similar to those of the monogastric stomach. In the abomasum, digestive juices containing acids and enzymes are added to aid in digestion.

Small intestine – In the first portion of the small intestine, called the duodenum, the digestive process started in the stomach continues. Nutrients are absorbed into the bloodstream through the walls of the remainder of the small intestine. Once the nutrients enters the bloodstream, they travel throughout the body to fuel life processes.

Large intestine – The large intestine consists of three parts, the cecum, colon, and rectum. Within the large intestine of the ruminant, the cecum plays a minor role in the further breakdown of roughage. The colon absorbs water and forms undigested wastes into feces, some absorption of nutrients also takes place. The feces are stored in the rectum before being passed out of the body.

Anus – The anus is the opening through which undigestible solid wastes exit from the body. Any portion of the feed not absorbed into the bloodstream is excreted through the anus.

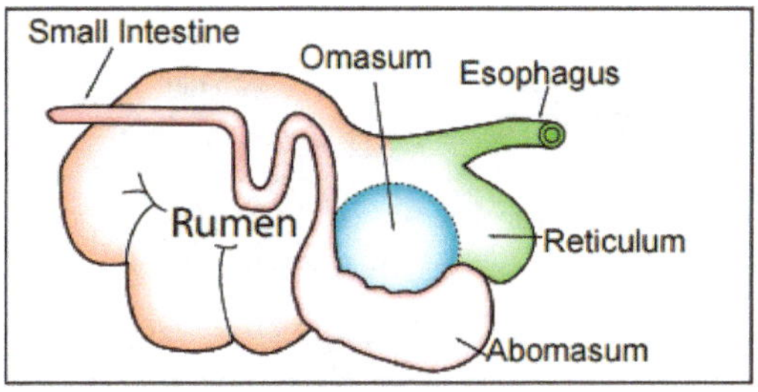

Figure - Ruminant stomach compartments.

The Digestive Process In ruminants, the breakdown of the nutrients in feed into a usable form is very complex. It involves not only the digestive juices and enzymes found in the monogastric system, but the activity of microorganisms found in the stomach. In addition, feed may take one of several paths as it passes through the stomach, depending on the nature of the feed.

Before feeds pass into the omasum and abomasum, they move through either the reticulum or the rumen and reticulum. Three types of microorganisms are present in the rumen and reticulum—bacteria, protozoa, and fungi. These microorganisms help break down feed for digestion. Bacteria have the most important role in digestion. Some bacteria break down the more easily digestible sugars and starches. Other bacteria ferment the fiber found in roughage like hay, grass, and silage to break it down for use by the body. A second type of microorganism, protozoa, also aids the ruminant in digestion by storing sugars and starches, forming some microbial proteins from proteins in the diet, and fermenting fiber. The role of fungi, the final type of microorganism, is not clearly understood, but they play a role in the digestion of fiber. The action of all the microorganism converts sugars, starches and fiber into fatty acids.

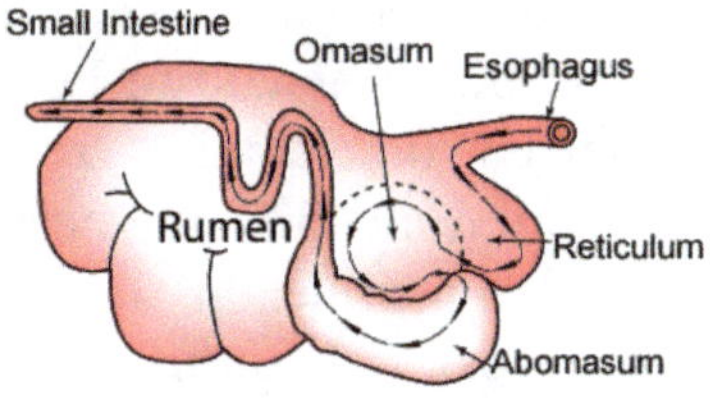

Ground Concentrates or Cud

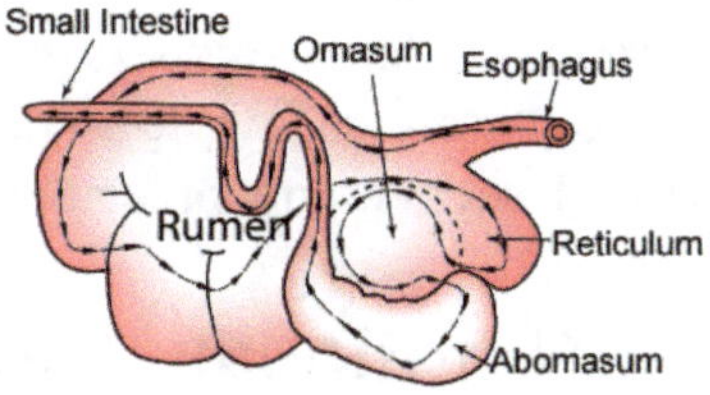

Light Grain

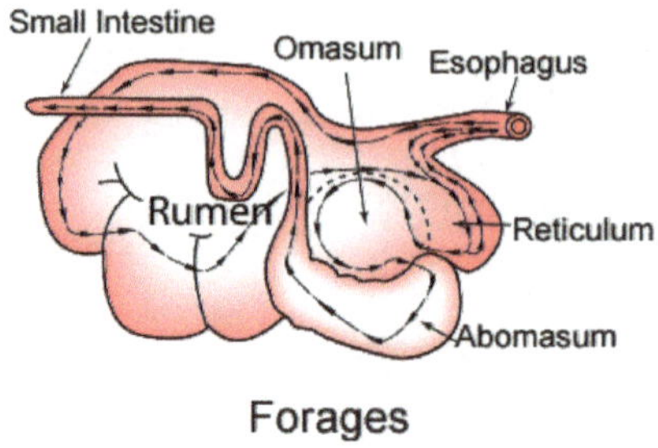

Figure Three possible routes of feed.

In the stomach, feed may follow one of three routes, depending on the type of feed. In the case of a ground concentrate or cud, the feed passes through the reticulum where the microorganisms act on it. It then passes through the omasum, where fatty acids are absorbed, to the abomasum. In the abomasum, gastric juices containing hydrochloric acid and the enzymes pepsin, rennin, and lipase are added to digest proteins and fats. Light grains, such as oats and barley, require a slightly more complicated path to be utilized most efficiently. They first pass into the rumen, in which muscular action breaks the feed into smaller pieces. The microorganisms then assist in digestion, and the fatty acids that are produced are absorbed by the rumen. When the grain is sufficiently broken down, it passes through the reticulum and omasum into the abomasum. The third route is that taken by forages. The combination of the action of the microorganisms and the path through the rumen and reticulum are what allows ruminants to better utilize forages, which have a high fiber content. As the forage passes into the rumen and reticulum, muscular action helps to break it down. The microorganisms then act on the pieces. After the microorganisms operate on the forage for a period of time, it moves to the reticulum. If the forage then needs to be broken down some more, the reticulum pumps it up through the esophagus into the mouth as cud to be chewed to break it into smaller pieces. After the cud has been chewed, the remains are again swallowed and enter the reticulum. They then pass to the omasum and abomasum. After passing through the abomasum, the feed enters the small intestine. In the duodenum, pancreatic juices are added, which include the enzymes trypsin, chymotrypsin, lipase, and amylase (of which only a small amount is present). As in the monogastric system, trypsin and chymotrypsin act on proteins, lipase on fats, and amylase on starches. Bile from the liver is also added in the duodenum for the digestion of fats. Intestinal juices are added in the rest of the small intestine. The enzymes peptidase, lactase, sucrase, and maltase are found in the intestinal juices; peptidase works on proteins, while maltase, sucrase, and lactase work on sugars and starches. Maltase and sucrase are present in small amounts in ruminants.

Importance of Meeting Nutritional Needs

Water and feed contain substances called nutrients that are needed by an animal for its body to function. The monogastric and ruminant digestive systems are able to extract these elements through digestion. To ensure proper growth and efficiency in production, it is important to provide the proper levels of nutrients for an animal's nutritional needs.

Nutrients

Nutrients are elements or chemical compounds that support the life processes of an animal. They allow the animal to maintain its bodily functions by entering the cells of the body and fueling their growth and function. Nutrients are found in water, grain, roughage, and other food substances taken into the digestive system through the mouth.

Six Basic Nutrients

Six basic nutrients are essential for the survival of the animal. They are water, proteins, carbohydrates, fats, vitamins, and minerals. The lack of any one of these nutrients could cause problems for the animal.

Water is a liquid combination of hydrogen and oxygen. Since the body is made mostly of water, water has many important functions.

Proteins are compounds made up of substances called amino acids that contain the elements carbon, hydrogen, oxygen, and nitrogen; iron, phosphorus, or sulfur are also sometimes included. Protein is the only source of nitrogen for an animal. The amino acids can be divided into essential and nonessential amino acids (Table 3.2). Essential amino acids must be provided in the diet because they are not found in animal tissues. Nonessential amino acids can be synthesized in animal tissues from other amino acids and therefore do not have to be supplied by the diet.

Protein has several functions. Protein is important because amino acids are the building blocks of body tissues. It develops and repairs body organs and tissues like muscles, nerves, skin, hair, hooves, and feathers. Protein is also used to produce milk, wool, and eggs. The fetus is developed and some enzymes and hormones are generated by the action of protein. It also forms a part of DNA.

Carbohydrates are made up of the elements carbon, hydrogen, and oxygen. A carbohydrate may be a sugar, starch, or fiber. Sugars and starches are easily digested, while fiber, which forms the cell wall material of plants, is more difficult to digest. The main function of carbohydrates is to provide the energy that powers muscle movement. Energy for muscular movement is required not only for exercise but for many essential functions of the body, such as breathing, digestion, and the beating of the heart. Carbohydrates also produce body heat, and extra carbohydrates are stored as body fat.

Fats, like carbohydrates, are an energy source. The elements found in fats are the same as carbohydrates – carbon, hydrogen, and oxygen. The difference between fats and carbohydrates is in the level of energy that they provide, with fats producing 2.25 times more energy than carbohydrates.

Table 3.2. Essential and nonessential amino acids

Essential Amino Acids	Nonessential Amino Acids
Arginine	Alanine
Histidine	Aspartic acid
Isoleucine	Citrulline
Leucine	Cysteine
Lysine	Cystine
Methionine	Glutamic acid
Phenylalanine	Glycine
Threonine	Lodogorgoic acid
Tryptophan	Proline
Valine	Serine
	Tyrosine
	Hydroxyglutamic acid
	Hydroxyproline

Minerals are inorganic elements that are utilized by the body. They include both macro- and microminerals; macrominerals are required in much larger amounts than microminerals. Minerals supply materials for building the skeleton and teeth and are a part of body regulators such as enzymes and hormones. They also assist in transmitting impulses through the nervous system, the development of the body tissues, and muscular activity. Table 3.3 provides a listing of the minerals needed by the body.

Vitamins are fat- or water-soluble organic substances. The fat soluble vitamins, carried by fats, are vitamin A, D, E, and K, which contain hydrogen, oxygen, and carbon. Water-soluble vitamins are vitamin C and the B-complex vitamins; they contain hydrogen, oxygen, carbon, chlorine, nitrogen, and cobalt or sulfur. Vitamins do not become a part of the body like the other nutrients. Instead, they regulate body functions. Vitamins regulate the digestion, absorption, and metabolism of nutrients. They also regulate the formation of new cells and the development of vision, bones, hair, feathers, skin and muscles. Vitamins help protect against diseases and develop and maintain the nervous system.

Nutritional Needs of Animals

Nutritional needs vary depending on the needs of the animal for health and the life stages of maintenance, conception and gestation, lactation, and growth and development. When a female is in the gestational period, for example, nutritional needs are different than during lactation. Figure demonstrates how the nutrient requirements of a mature female vary through different life stages. Of the life stages, simple maintenance has the lowest level of nutritional requirements, and they must be met before other nutrients needs are considered. At this level, nutrients are used to maintain vital life processes and normal body temperature, with no weight gain or loss and no production (such as reproduction or fattening). The amount of feed needed to maintain an animal depends on the size of its body. The main concern during conception and gestation is meeting additional nutritional needs to ensure good fetal growth and maintain the health of the mother. Most of the growing done by the fetus is done in the last trimester.

Table 3.3. Macro- and Microminerals

Macrominerals	Microminerals
Calcium (Ca)	Iron (Fe)
Phosphorus (P)	Iodine (I)
Sodium (Na)	Copper (Cu)
Potassium (K)	Cobalt (Co)
Chlorine (Cl)	Fluorine (Fl)
Magnesium (Mg)	Manganese (Mn)
Sulfur (S)	Zinc (Zn)
	Molybdenum (Mo)
	Selenium (Se)
	Chromium (Cr)

During this time the female needs additional nutrients for herself and the fetus. The amount needed is equal to the nutrient needs of the young animal after birth and the maintenance needs of the female.

Lactation is milk production. Many additional nutrients are needed to produce milk, including proteins, minerals, vitamins, fats, and carbohydrates. Protein is important, since milk contains three percent protein. The energy supplied by fats and carbohydrates is vital for lactation. Water is also needed. Two minerals needed during lactation are calcium and phosphorus.

An animal also has special nutrient requirements for growth and development. As it grows, the animal increases the number of cells at the tissue level. Muscles are built, and bone and connective tissues are produced. To accomplish this task, an increased amount of protein is needed. The minerals calcium and phosphorus are also required for bone growth. In addition, higher quantities of vitamins are required, especially vitamin D. An energy source is supplied by including more fats and carbohydrates in the animal's diet. If these nutrient requirements are not met while the animal is growing, the adult will be less productive.

Nutrients are important in maintaining the health or well-being of the animal in all of the life stages. Nutrients are needed to carry out vital bodily functions.

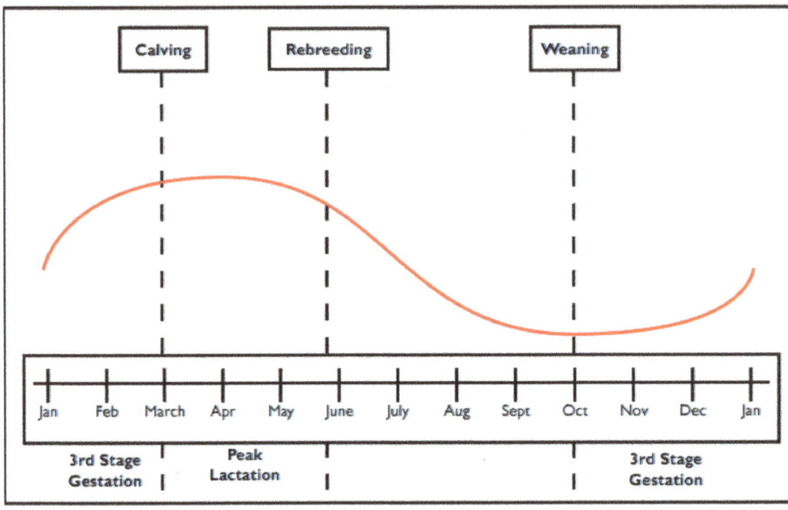

Figure : Nutrient Requirements for a cow through life stages.

MEETING NUTRITIONAL NEEDS OF ANIMALS

In order to meet the animal's nutritional needs, the nutrients in the feed have to match the animal's requirements.

Information Sources for Nutritional Requirements

A number of sources provide information that the producer can use to help him or her meet an animal's nutritional needs.

For example, University Extension has livestock and dairy specialists and agricultural publications that are good information sources on nutritional needs. Extension guides that list nutritional information based on the animal's life stage, size, and weight are available for many animal species.

Steps in Meeting Nutritional Needs of Animals

Some general steps should be followed in order to meet the nutritional needs of animals. The goal of the step-by-step process is to meet the animal's needs through the nutrients provided in the feed.

When trying to meet nutritional needs, the animal and the type of feed have to be taken into consideration.

Research needs to be done to learn the nutritional needs of the animal and the nutrient content of feeds. The first step is to find one of the information sources listed above. The source can then be used to obtain information on the requirements of that animal. Information on feeds may sometimes be found in the same source that provides data about the nutritional requirements of an animal. If information on feeds is not included, however, another source must be found.

After a source has been obtained, the nutritional needs of a particular animal need to be identified. This step has several parts. First, the life stage of the animal has to be identified.

Using this information, the producer can use the source to look up all the nutrients needed by the animal. Table 3.4 shows a section of a table that supplies information on nutrient requirements.

The next step in meeting an animal's nutritional needs is to identify the nutrients that are available in a specific feed. This part of the process is as important as identifying the animal's needs, since the feed must supply the required nutrients. This step also has several parts. First, the feeds that may be used must be identified. Many feeds may be available to the producer, such as corn, soybean meal, grass, and hay. Next, a version of a particular feed should be chosen to meet the animal's nutritional needs. For example, whole seed soybeans have a different amount of nutrients than dehulled soybean meal. The producer needs to pick the right one for his or her purposes. In order to pick the appropriate feed, the producer has to know what level of nutrients is provided by that specific feed, which can be learned from the source used. Table 3.5 shows a portion of a sample feed nutrient composition table.

Table.3.4. Nutrient Requirements Table

Weight (lbs.)	ADG (lbs.)	DM[b] (lbs.)	Protein (lbs.)	Ne_m Mcal	Ne_g Mcal	TDN (lbs.)	Ca (lbs.)	P (lbs.)
Medium-Frame Steer Calves								
400	2.5	11.0	1.56	3.81	3.16	8.1	.074	.033
400	3.0	10.0	1.65	3.81	3.86	8.5	.086	.037
500	0.5	11.5	0.98	4.50	0.64	6.2	.028	.019
500	1.0	12.3	1.16	4.50	1.37	7.2	.039	.024
500	1.5	12.8	1.33	4.50	2.13	8.1	.051	.028
500	2.0	13.1	1.49	4.50	2.93	8.8	.061	.031
500	2.5	13.0	1.63	4.50	3.74	9.6	.072	.035
500	3.0	11.8	1.69	4.50	4.56	10.0	.081	.037
600	.05	13.2	1.08	5.16	0.73	7.1	.030	.023
600	1.0	14.1	1.26	5.16	1.57	8.2	.039	.026

The final step is to match the feed to the animal's nutritional requirements to meet its needs.

Table 3.5. Feed Composition Table

	Dry Matter %	Crude Protein%	Nem Mcal/lb	Neg Mcal/lb	TDN %	Fat %	Crude Fiber %	Calcium %	Phos-phorus %	Potas-ium %	Sodium %	Sulfur %	Zinc PPM
Concentrates													
Molasses													
Beet	78.0	6.6	0.68	0.45	62	0.2	0.0	0.13	0.02	4.73	1.15	0.47	14
Cane	75.0	4.4	0.58	.037	54	0.1	0.0	0.75	0.08	2.88	0.17	0.35	23
Cane, dried	94.0	9.7	0.70	0.44	66	0.8	6.3	1.03	0.14	3.38	0.19	0.43	31
Oats													
Grain	89.0	11.8	0.75	0.49	69	4.8	10.8	0.06	0.34	0.39	0.07	0.20	37
Mill Byproduct	89.0	7.1	0.34	0.00	29	2.3	22.3	0.11	0.21	0.53	-.-	0.21	-.-
Groats	90.0	15.9	0.95	0.65	85	6.2	2.5	0.07	0.43	0.35	0.05	0.20	0

- Protein – Amino acids found in animal tissues, grains, and legumes
- Carbohydrates – Sugars, starches, or fibers found in grains and forages
- Fats – Fatty acids found in animal tissues, fats, and grains
- Vitamins – Found in grass, sun-cured hay, and commercial feeds
- Minerals – Found in plant and animal tissues and commercial feeds

BALANCING RATION FOR ANIMAL

Once a producer has determined the nutritional needs of an animal and the nutrient content of feeds, the next step is balancing a ration. A balanced ration provides the needed nutrients for the animal by including the necessary amounts of the proper feeds.

Importance of Balanced Rations

A balanced ration provides the proper levels of nutrients for an animal. The objective of balancing a ration is to meet the animal's nutritional needs with feed that supplies each nutrient in the proper amount and proportion. See Figure . Providing animals with balanced rations is important. Balancing rations will allow a producer to meet an animal's nutritional needs for health and the specific nutrient requirements of each life stage—maintenance, conception and gestation, lactation, and growth and development. By balancing rations, nutritional deficiencies that can harm the animal can be prevented. In addition, growth and production can be maximized.

Figure - Balancing Nutrients in Rations.

Methods of Balancing Rations

Several methods may be used to balance a ration. Special computer programs, trial and error, and the Pearson square method may all be utilized by the producer when balancing a ration. In addition to these common methods, many other methods are available to producers. Today, computers are widely used when balancing rations. Local feed dealers, cooperatives, companies, and producers may balance a ration using computers. Special programs have been designed to check that the feeds included in a ration meet all of the animal's nutrient requirements by making sure they are balanced. The computer does the figuring with the information provided about the animal and the type and amount of feed. It therefore requires complete information for accurate results. When this data is entered, the program will indicate whether the ration is balanced, the animal needs more of a certain nutrient or nutrients, or the ration is exceeding the animal's needs. If the ration is not balanced, the person using the program must alter the ration and enter the new information to allow the computer to check it.

Another method for balancing rations is referred to as the trial and error method. To use this method, a combination of feeds is selected that the producer believes will meet nutritional needs. The nutrient content of the diet is calculated and compared to the nutritional requirements of the animal. If the diet does not meet those requirements, a different combination of feeds must be selected and the process repeated. The producer attempts to meet all the animal's nutrient needs with this method.

A third method, the Pearson Square method, can also be used to balance rations. This method balances for only one nutrient at a time; it is most often used for crude protein. In order to balance for other nutrients, a new Pearson square must be constructed. This method is sometimes utilized by a producer engaged in limited production. It is a good way to start learning how to balance rations.

The process for using the Pearson Square is quite simple. First, draw a square similar to that in Figure . Write the percent of the nutrient required by the animal in its diet in the middle of the square.

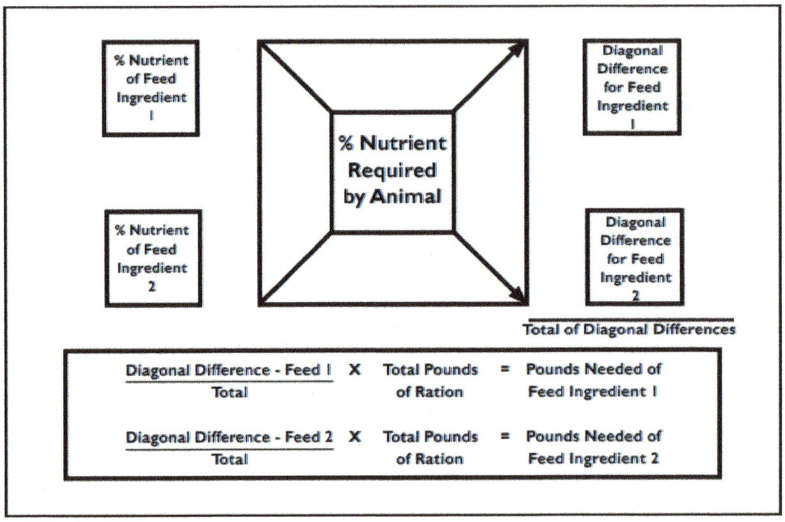

Figure . Pearson square method.

Find the difference by subtracting diagonally through the square; always subtract the smaller number from the larger number. Write the results on the right corners of the square. The answer in the upper right corner represents the proportion of the ingredient in the upper left corner that will be included in the ration, while the answer in the lower right corresponds to the proportion of the ingredient on the lower left that is included. Next add the numbers in the right-hand corners and place a total at the bottom. Once the square has been completed, some additional calculations must be done to determine the amounts and proportions of feeds needed. First, a percentage is found for the ingredients by taking the diagonal difference for each ingredient and dividing it by the total. This number is the percentage of that ingredient needed in the ration in order for the ration to balance. After the percentage of each feed ingredient is found, the number of pounds of feed can be determined by multiplying the percentage by the total number of pounds in the ration. After the ration is balanced for one nutrient using this method, the producer must still check to ensure that the proper amounts of other nutrients are supplied to the animal.

Factors Influencing Ration Ingredients

Many different feeds can be used to meet an animal's nutritional requirements. A combination of feeds can also be used. The selection of a feed or combination of feeds for the animal is influenced by two factors, least cost and ingredient availability.

The term "least cost" refers to getting the desired performance at the least cost to the producer. The type of nutrients needed by the animal cannot be ignored on the basis of cost, however. Suppose corn is $2 a bushel, while wheat can be bought for $4 a bushel. Which one should be fed? If the corn is chosen because it is cheaper and corn does not meet the nutritional needs of the animal, then in the long run costs may actually be greater, since growth and production are not maximized.

The process for determining least cost is not complicated. First, each feed ingredient in a ration has to be balanced. Once the amount of the particular feed ingredient is known, the total cost of the ration can be calculated by multiplying the quantity of each feed by its price and then adding to find total cost. The costs of balanced rations using different feed ingredients can then be compared. Choosing one ration over another ration that supplies the same nutrients to the animal will depend on which ration has the least cost.

The other factor that influences feed selection is ingredient availability. The availability of a particular feed depends on the geographical area. Areas vary in the types of feeds that are available. For example, in places where grains are grown extensively, corn will be more available as a feed ingredient.

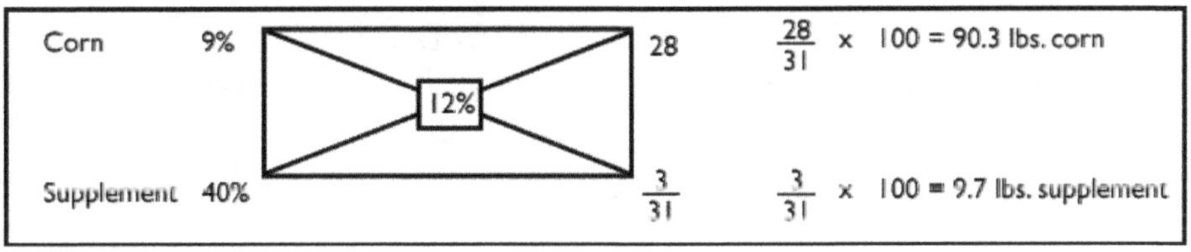

Figure . Sample Pearson square calculation.

Areas with an abundance of good grazing, on the other hand, will utilize forages for feed. In addition, different types of manufacturing may produce byproducts that are available as feed to animals. Manufacturing byproducts that are fed to animals include items such as fish meal, blood meal, brewer's grains, whole cottonseeds, and soyhulls. Ingredient availability therefore depends on what is grown or manufactured in a certain area.

REFERENCES

1. Butler W. R., R. D. Smith. 1989. Interrelationships between Energy-Balance and Postpartum Reproductive Function in Dairy Cattle. J. Dairy Sci. 72:767-783.
2. Reist M., D. Erdin, D. von Euw, K. Tschuemperlin, H. Leuenberger, Y. Chillard, H. M. Hammon, C. Morel, C. Philiponaa, Y. Zbinden, N. Kuenzi, J. W. Blum. 2002. Estimation of energy balance at the individual and herd level using blood and milk traits in high-yielding dairy cows. J. Dairy Sci. 85:3314-3327.
3. LeBlanc S., 2010. Monitoring Metabolic Health of Dairy Cattle in the Transition Period. J. Reprod. Dev 56:S29-S35.

4. Oetzel G. R., B. E. Miller. 2012. Effect of Oral calcium bolus supplementation on early-lactation health and milk yield in commercial dairy herds. J. Dairy Sci. 95:7051-7065.
5. McArt J. A. A., D. V. Nydam, P. A. Ospina, G. R. Oetzel. 2013. A field trial on the effect of propylene glycol on milk yield and resolution of ketosis in fresh cows diagnosed with subclinical ketosis. J. Dairy Sci. 94:6011-6020.
6. LeBlanc S. J., K. E. Leslie, T. F. Duffield. 2005. Metabolic Predictors of Displaced Abomasum in Dairy Cattle. J. Dairy Sci. 88:159-170.
7. McArt et al ., 2012
8. Duffield, T. F., K. D. Lissemore, B. W. McBride, K. E. Leslie. 2009. Impact of hyperketonemia in early lactation dairy cows on health and production. J. Dairy 92:571-580.
9. Chapinal N., S. J. LeBlanc, M. E. Carson, K. E. Godden, M. Capel, J. E. P. Santos, M. W. Overton, T. F. Duffield. 2012. Herd-level association of serum metabolites in the transition period with disease, milk production, and early lactation reproductive performance. J. Dairy Sci. 95:5676-5682
10. McLaren C. J., K. D. Lissemore, T. F. Duffield, K. E. Leslie, D. F. Kelton, B. Grexton. 2006. The relationship between herd level disease incidence and a return over feed index in Ontario dairy herds. Can. Vet. J. 47:767-773.ss

CHAPTER 3

Cattle Health

Chapter Contents

Maintaining animal health is the key to a profitable and satisfying animal enterprise. There are several considerations that need to be made in dealing with the health of animals. These include being able to recognize signs of good and poor health; maintaining a healthy environment; being able to identify animal diseases and parasites; and knowing how to treat health problems when they occur. Diseases are infective agents that result in lowered health in living things. Parasites are animals that live on other animals and derive their food from their hosts.

INTRODUCTION

Everyone gets sick from time to time, and animals are no different. But how do farmers keep animals healthy and when they are sick, make them better? The most important things a farmer can provide animals are lots of clean water, feed and a healthy environment. These simple, yet crucial, items allow the animals to behave naturally and go through natural processes such as eating, resting, growing and, for some animals, making milk.

There are many different ways that farmers help to prevent sickness and injury on their farms. One way is for farmers to vaccinate their animals against diseases, very similar to the way humans are vaccinated. Farmers work with veterinarians to develop vaccination programs that outline which vaccinations to give and at what age each is to be given.

SIGNS OF HEALTH AND ILL HEALTH IN ANIMALS

When you have a farm animal and you don't know what causes it to get sick all the time then here are some signs you need to take notice of if it is in good health or ill health. The health of the animal shows whether if you are feeding it well enough and keeping its feed and environment clean. Or if you are giving it less feed which is deficient in certain nutrient and might have been contaminated with foreign materials which is causing the dysfunction of its internal organs and external organs.

- **Appearance and behavior:** Healthy animals exhibit normal posture. They are alert, having clear eyes and response to touch. Any abnormal running nose and dull eyes may indicate ill health
- **Movement:** Any animal that cannot walk or stand properly must be closely watched since something might be wrong with it. When you notice that the animal is walking and favoring on leg it will be very advisable if you restrain that animal and find out what is causing it to do so.
- **Appetite and feeding:** Healthy animals have good appetite for food and generally like to feed to their satisfaction. Sick animals have no appetite for food. When you notice that your animal suddenly lose appetite for food you give it, you shouldn't take it that it is normal because it may be a sign of it beginning to fall ill.
- **Urine:** The normal color of urine is pale yellow. Much deviation like deep yellow, blood stained or cloudy urine shows ill health. When your animal show pain urinating shows that there is something wrong with its urinal system and any other color a part from pale yellow is a sign of ill health.
- **Coughing:** when an animal cough continuously, it shows that something is irritating its throat and you need to find out why that keeps happening. Coughing is a good sign of good health but it shouldn't be continues cough.
- **Pain:** Healthy animals do not show any sign of pain but when animals begins to show signs of pain by grinding of teeth or groaning, it is an indication that something is wrong with it and need to be examined immediately.
- **Faeces and defecation:** Any deviation of the faeces i.e. too hard, too watery or stained with blood or contaminated with worm segments, is an indication of ill health. When you see that your animal begins to defecate on its body it is an indication that it has a problem in its alimentary canal and immediate attention need to be given to the animal.
- **Skin and coat:** The coat of healthy animals should be clean, smooth and shiny and show complete cover. Signs of ill health are; the coat looks dull and hairs fall out. Cold, dry and scary skin signifies diseases. When you see your animal suddenly show symptoms of skin disorder, immediate attention should be given

to the animal because when you don't your animal hide reduces in value.

- **Temperature, pulse rate and respiratory rate:** Every breed of farm has a specific range of body temperature, pule rate and respiratory rate that considered normal for the healthy animals. Radical deviation from these values signifies ill health. You will have to check these of your farm animal to ensure that it is physically, mentally and psychologically fit. Any deviations in these shows that your farm animal is ill.
- **Mucous membrane:** These are the lining of various system and organs of the body, e.g. eyes, nose, anus etc. The mucous lining of healthy animals is moist. When all these places become dry the animal experiences discomfort and when noticed immediate attention should be given to the animal before it becomes very severe.

Signs of Health in farm animals

Healthy animals like people should look as if they are enjoying life, Here are points to look out for:

Healthy Animals	Sick Animals
They have close contact although scattered will run towards each other to form a group.	Usually isolated and not running with the flock/herd.
Definite physical activity. They are usually eating, walking, playing, grooming or cuddling. They will always stretch after rising.	Immobile and not interested.
Farm animals tend to be greedy if they are well and always want food.	Ill animals have no/little appetite.
The excretion pattern is the same and the dung consistency matches the feed type. There is never any blood and the urine is a light colour.	Diarrhoea or the absence of dung or urine is a sure sign that there is something wrong.
The bovine coat is usually shiny and has lick marks on it. The sheep's fleece is tight and not matted and it should be springy.	A dull coat in a bovine is a bad sign. If sheep are ill, usually their wool breaks or falls out.
The eyes are clear, bright and alert. Eye sockets should be pinky.	Ill animals have sunken eyes and dull. Eyelids stop blinking. If eye sockets are pale/white, this usually means anaemia.
There are no abnormal discharges from the body.	Ill animals can have discharges from eyes, mouth, nose, navel, anus, vagina, teats, ears or penis.

Healthy animals walk well and in a straight line.	Ill animals can walk stiffly, wobble or go round in a circle. Creaking and swollen joints are signs of ill health.
The metabolic traits (pulse rate, breathing and heartbeat) are normal.	The metabolic traits are higher, lower than normal or are erratic.

The importance of taking temperature

The temperature of a farm animal can be used to diagnose ill health according to a pathogen.

Animal	Temperature Degrees Centigrade	Respiration rate Per minute	Pulse rate Per minute
Cattle	38.7	12-20	45-50

Recognizing Ill Health in Animals

There are degrees of ill health ranging from the animal that is merely "off-color" to one that is desperately ill. An animal that looks "not quite right" should be observed closely until it appears fully recovered. If it is incubating a serious disease, an early diagnosis could save the animal. By checking the vital signs of the animal, the owner can receive early warning that something is amiss. Seriously ill animals must receive immediate and urgent veterinary attention.

The first sign that an animal is becoming sick is that it picks at or refuses food. It may drink more or less water than normal, depending on the illness. The eyes will be dull, and on closer inspection, the mucous membranes may have changed color. Deep red membranes indicate fever; pale membranes show anaemia; yellow membranes indicate a liver disorder, while blue-red membranes show heart and circulatory problems, or pneumonia.

The coat may look dull and dry. The animal might be sweating (except for dogs). A cold sweat indicates pain while a hot sweat indicates fever. If the animal is in pain it will probably be restless (getting up and down and pacing about) and it might even be groaning.

The animal will either scour (i.e. pass very loose droppings), or will become constipated and pass no droppings at all. The passing of urine might also cease. A very sick animal will lie down for long periods and will not get up when approached.

The vital signs of a sick animal will change. The temperature may go up or down. A rise in temperature of one or two degrees usually indicates pain, while a rise of more usually indicates infection.

The rate of respiration, and the way the animal breathes could also show changes. With pain or infection, breathing becomes more rapid. In a very sick animal, breathing can be labored and shallow.

A slightly increased pulse rate suggests pain, while a rapid pulse suggests fever. An irregular pulse can indicate heart trouble. In a very sick animal, the pulse is weak and feeble.

Animal owners should consistently check pets for signs of health or ill-health. Aside from the above vital signs, the following should be monitored:

- General condition – weight and body condition
- Sluggishness
- Movement or Gait – check for lameness
- Appetite
- Excessive Thirst
- Weight Loss
- Eyes
- Ears
- Mouth
- Coat
- Skin
- Head
- Teeth
- Limbs
- Trunk
- Mucus membranes
- Swellings
- Excitability
- Slobbering
- Discharge
- Urination
- Diarrhoea
- Breathing
- Vomiting

APPEARANCE OF THE HEALTHY ANIMAL

The healthy animal is alert and aware of its surroundings. It is active and holds its head up watching what is happening around it. It should stand on all of its feet. The separation of an animal from the others in its group is often a sign of a health problem.

An animal which is not interested in its surroundings and does not want to move has health problems.

Movement (gait)

The healthy animal will walk easily and steadily with all of its feet taking its weight. Steps should be regular. Irregular movement results from pain in the feet or limbs.

Horses normally stand during the day. If you go near an animal that is lying down it should stand up quickly otherwise it has health problems.

Eyes

The eyes should be bright and alert with no discharge at the corners.

Ears

Most animals have erect ears which move in the direction of any sound. Ear movements will also be quick to get rid of flies, his body temperature of the pig can be checked by touching the ear when an unusually high temperature will be noticed.

Nose and Muzzle

The nose should be clean with no discharge. In cattle and buffalo the muzzle should be moist not dry. In sheep and goats the nose should be cool and dry. Healthy animals frequently lick their noses with their tongues.

Nose and Muzzle

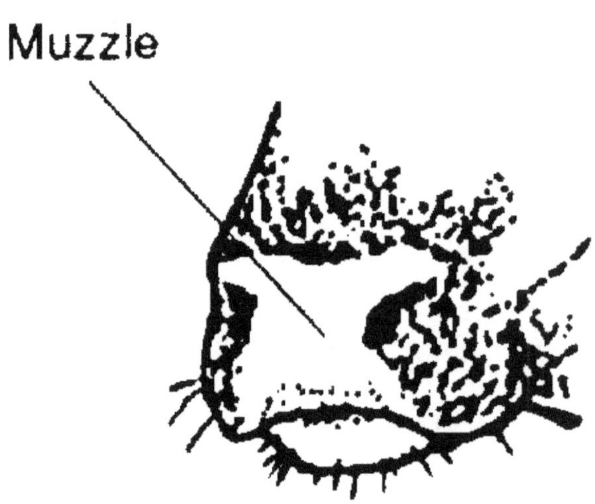

Mouth

There should be no saliva dripping from the mouth. If chewing is slow or incomplete there must be a problem with the teeth.

The Coat

In short-haired animals, e.g. goat and cattle, the hair or coat of the healthy animal will be smooth and shiny. Healthy cattle, buffalo and their calves lick their coat and the lick marks will show. Horses should not sweat when resting.

The coat

In poultry the feathers should be smooth and glossy and not ruffled. In pigs a curly tail is a sign of good health while a scaly skin points to health problems.

Behavior

If a horse, cow or buffalo keeps looking at its flanks or kicks at its belly it has a pain in the stomach.

Breathing

Breathing should be smooth and regular at rest. Remember that movement and hot weather will increase the rate of breathing. If the animal is resting in the shade it should be difficult to notice the chest moving as it breathes.

Pulse

Taking the pulse is important when examining an animal. In man the pulse can be easily taken but in animals it is more difficult and requires practice.

- In sheep and goats you can feel the pulse on the inside of the top of the back leg. The rate of the pulse is 70 - 130 per minute in the adult.
- The pulse of cattle is taken at a point on the underside of the base of the tail, the normal rate is 40 - 80 per minute in the adult. In buffalo the pulse rate is 40 - 60 per minute.

- The pulse of the horse is taken on the inside of the cheek. The normal rate is 35 - 40 per minute.
- The pulse of the camel is taken at a point on the underside of the root of the tail. The normal rate is 35 - 45 beats per minute.

In the llama, alpaca and the pig there is no point at which the pulse can be taken. In these animals the beat of the heart itself must be felt for.

Droppings or Dung

The droppings of the healthy animal will be firm. Very soft droppings (diarrhoea) is a sign of ill health. If the animal has difficulty in defecating (constipation) this is also a bad health sign.

Urine

The urine should be clear and the animal show no signs of pain or difficulty in urinating. Horses, mules and donkeys can have thick yellow urine which is normal.

Appetite and Rumination

The animal should eat and drink normally. Failure to eat is an obvious sign of ill health. If feed is available the healthy animal will have a full belly. Pigs will naturally rush at their feed, if they do not something is wrong. Sheep, goats, cattle, buffalo and camels chew the cud (ruminate) for 6 to 8 hours each day. It is a sign of ill health when these animals stop ruminating.

Milk

In the milking animal a sudden change in the amount of milk produced can mean a health problem. Any sign of blood or other matter in the milk points to infection in the udder. There should be no swelling of the udder and no sign of pain when it is touched. There should be no injury to the teat.

Spread of Disease

Disease can be classified as acute or chronic. An acute disease starts quickly and lasts for a short period when the animal either recovers or dies. A chronic disease lasts for a long time and weakens the animal. Diseases are said to be infectious (will spread from one animal to another) or noninfectious (will not spread from one animal to another).

Non-infectious diseases can be caused by poor feed and the lack of minerals, salts and vitamins that the body needs. Non-infectious disease can also be caused by poisoning with chemicals or plants, by cuts, burns and broken bones. Some diseases pass from the parent to the young (hereditary).

Many non-infectious diseases are chronic but they can be acute. They can cause large losses of meat, milk and wool. Working (draught) animals do not work well and the rate of reproduction can be low with the young being born dead or dying before they are weaned. Chronic diseases are often thought to be "normal" but when the cause is known and eliminated production can be greatly increased.

Infectious diseases are caused when the body is attacked by tiny living germs.

The Spread of Disease

Infectious diseases can be spread by:

- Direct contact between animals.
- Germs in feed and water.
- By faeces and urine from sick animals.
- By flies, ticks, lice and fleas.
- By dirty housing or shelters.
- Young and old animals become infected more easily.

Preventing Infectious Diseases

- Animals, like humans, must be clean in order to be healthy. The animal must be provided with clean feed, water, bedding and shelter.
- Sick animals should be kept separate from the others.
- Some diseases can be cured by drugs.
- Vaccination can protect animals against some diseases.
- Dead animals and waste should be disposed of.

Preventing Non-Infectious Diseases

The chronic non-infectious disease may not be recognized as a disease. The affected animals may not die but will not produce as much milk, meat or wool, or work as well as could be expected.

If we continually look for ways of improving feed, water, mineral and vitamin supplies we will find the way to control the non-infectious diseases. This will lead to greater production of wool, meat and milk, draught animals will be stronger and more young will be produced. Poultry will produce more eggs and get fatter.

HEALTHFUL ENVIRONMENTS FOR ANIMALS

Maintaining a healthy environment for animals is a key factor in a complete animal-health program. It is often much less expensive to maintain a healthy environment for animals than it is to treat animals that are unhealthy due to poor conditions that occur.

Sanitation

Good sanitation is important to good health. Factors related to good sanitation include keeping facilities for animals clean. Sanitation also requires the use of clean equipment when dealing with animals. This includes feed containers; milking equipment; artificial-breeding equipment; needles and syringes; and surgical equipment. A syringe is an instrument used to give injections of medicine or to draw body fluids from animals (Figure .1). Simple on-farm surgical procedures should always be performed with the strictest sanitation possible. The liberal use of disinfectants in dealing with animals is also important. A disinfectant is a material that kills disease-causing organisms.

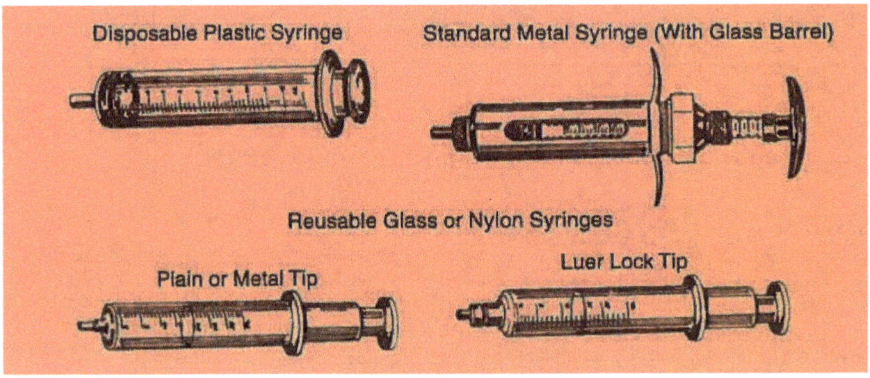

Figure .1: Types of syringes.

Housing

Maintenance of proper housing is an important consideration in maintaining good animal health. Housing should be clean and free from cold drafts. However, good air circulation throughout the housing is important to help lower high temperatures in the summer and reduce humidity in the cold of winter. Extremely dry and dusty conditions are to be avoided whenever possible. Proper maintenance of animal housing is important too. Loose boards, roofing materials, and nails often pose problems in poorly maintained facilities.

Handling Manure

Piles of manure, dirty pens, and dirty feedlots are often sources of serious health problems in animals. It is important that manure not be al-lowed to accumulate in areas frequented by animals (Figure .2). Manure piles often harbor diseases and parasites. They also attract flies, which may spread diseases. Cages and pens soiled continually with animal waste products may also lower the quality of the air breathed by the animals. Wet, poorly drained, manure-soiled feedlots usually re-duce the rate of gain in beef cattle and swine. Feedlots are areas in which large numbers of animals are grown for food. Feet and leg problems can often be traced to poorly maintained feedlots.

Figure .2: An efficient manure handling disease and prevent parasites.

Controlling Pests

The control of pests and parasites is also an important consideration in the maintenance of animal health and welfare. Regular use of disinfectants to control parasites such as lice and flies is necessary in a good disease-prevention program. Regular, close observation of animals may be necessary to determine when outbreaks of parasites occur. Prevention of such parasites is preferable to controlling outbreaks that occur. To that end, the development of a good prevention program is a wise decision (Figure .3). The control of other pests, such as birds and wild animals, is also part of a good animal-health program. Many birds carry parasites on their bodies and in their droppings. When the parasites move from infected animals to healthy ones, they often carry diseases and parasites with them. Wild animals and pets may also cause serious health problems when allowed to roam freely around farm animals. Dogs and coyotes will often chase animals and cause them to injure themselves. Bites from these animals may also cause infection and other health problems. Just the presence of pets around farm animals may cause the farm animals to be nervous and may affect how they grow and produce.

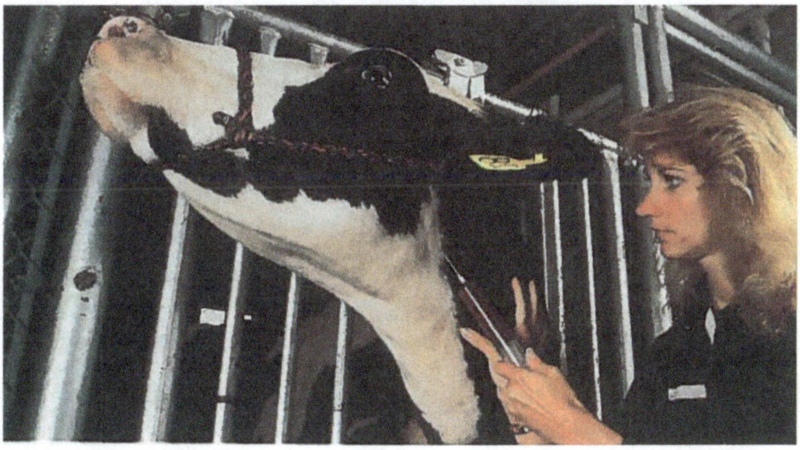

Figure .3: The drawing and analyzing of blood samples is a techniques for diagnosing and treating many diseases and disorders in animals.

Isolation

The isolation of animals new to a herd is an important part of any good health-prevention program. Such animals may be harboring disease or parasites that are not readily apparent. It is wise to keep them isolated from other animals for a period of time, usually a minimum of thirty days. This gives the new owner time to observe the isolated animals closely for health problems. Similarly, isolation of diseased animals is important. Animals with contagious diseases that can be spread by contact should never be al-lowed to come into contact with healthy animals. It is difficult to treat unhealthy animals when they are living with large groups of animals. Healthy animals tend to pick on unhealthy ones, making it especially difficult for such animals to regain health.

Pasture Rotation

The rotation of pastures is a consideration in maintaining a healthy environment for animals and in preventing health problems. Many diseases of animals are harbored in the soil and are killed only by not being able to come into contact with host animals for extended periods of time. A host animal is an animal in or on which diseases or parasites can live. Moving animals to different pastures on a regular basis also allows for the breakdown of animal wastes and for pasture regrowth.

Figure .4: Pasture rotation is necessary to break the life cycles of diseases, insects, and internal para-sites, and pays additional dividends in greater pasture production.

ANIMAL DISEASES

Animal disease an impairment of the normal state of an animal that interrupts or modifies its vital functions. Concern with diseases that afflict animals dates from the earliest human contacts with animals and is reflected in early views of religion and magic. Diseases of animals remain a concern principally because of the economic losses they cause and the possible transmission of the causative agents to humans. The branch of medicine called veterinary medicine deals with the study, prevention, and treatment of diseases not only in domesticated animals but also in wild animals and in animals used in scientific research. The prevention, control, and eradication of diseases of economically important animals are agricultural concerns. Programs for the control of diseases communicable from animals to man, called zoonoses, especially those in pets and in wildlife, are closely related to human health. Further, the diseases of animals are of increasing importance, for a primary public-health problem throughout the world is animal-protein deficiency in the diet of humans. Indeed, both the United Nations Food and Agricultural Organization (FAO) and the World Health Organization (WHO) have been attempting to solve the problem of protein deficits in a world whose human population is rapidly expanding.

Economic Importance

About 50 % of the world's population suffers from chronic malnutrition and hunger. Inadequate diet claims many thousands of lives each day. When the lack of adequate food to meet present needs for an estimated world population of more than 4,600,000,000 in the 1980s is coupled with the prediction that the population may increase to 7,000,000,000 by the year 2000, it becomes obvious that animal-food supplies must be increased. One way in which this might be accomplished is by learning to control the diseases that afflict animals throughout the world, especially in the developing nations of Asia and Africa, where the population is expanding most rapidly. Most of the information concerning animal diseases, however, applies to domesticated animals such as pigs, cattle, and sheep, which are relatively unimportant as food sources in these nations. Remarkably little is known of the diseases of the goat, the water buffalo, the camel, the elephant, the yak, the llama, or the alpaca; all are domesticated animals upon which the economies of many developing countries depend. It is in these countries that increased animal production resulting from the development of methods for the control and eradication of diseases affecting these animals is most urgently needed.

Despite the development of various effective methods of disease control, substantial quantities of meat and milk are lost each year throughout the world. In countries in which animal-disease control is not yet adequately developed, the loss of animal protein from disease is about 30 to 40 % of the quantity available in certain underdeveloped areas. In addition, such countries also suffer losses resulting from poor husbandry practices.

ANIMALS PARASITES

Parasites may also be grouped into two general classifications. These are internal—inside the animal—and external—living on the outside of the animal. The most important group of internal parasites that infest animals are the roundworms (slender worms that are tapered on both ends). Other types of internal parasites include flukes and protozoa. Flukes are very small, flat worms, and protozoa are microscopic, one-celled animals. Most internal parasites spend at least some of their life cycles outside of the host animal (Figure .5). It is during this period that the parasite may most easily be spread to other animals. Contact with dis-charges from infested animals, contaminated feed, water, housing, or secondary hosts may result in the spread of internal parasites. A secondary host is a plant or animal that carries a disease or parasite during part of the life cycle of that disease or parasite. Some internal para-sites are also spread by insects such as flies and mosquitoes.

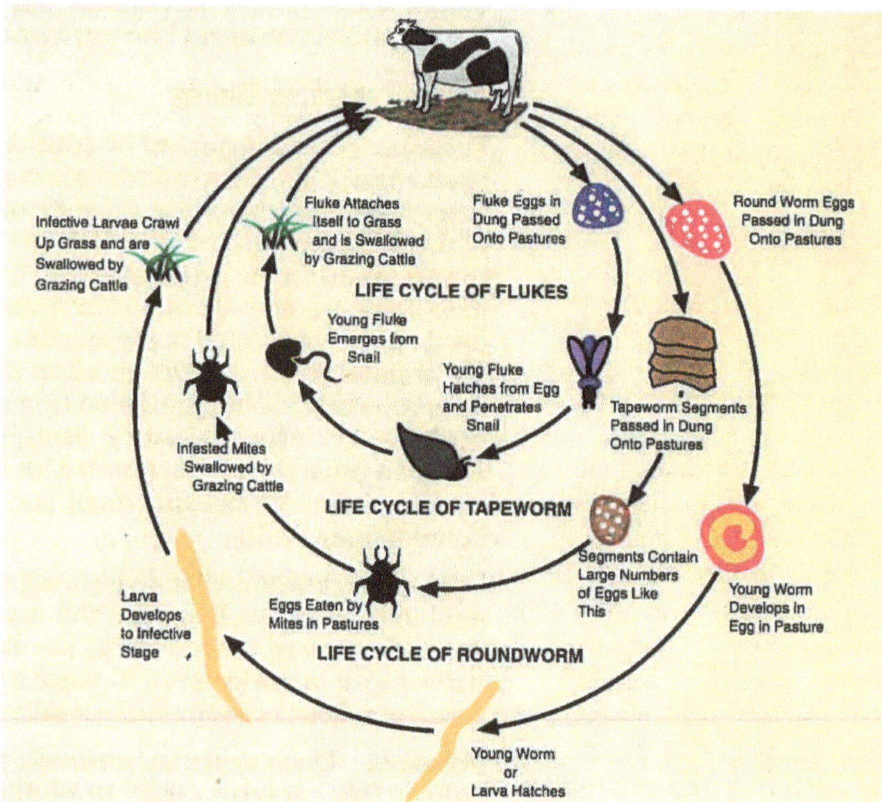

Figure .5: Life cycles of some common internal parasites.

External parasites include flies, ticks, mites, and fleas. They are spread in the same ways as are internal parasites. Symptoms of parasite infestation may include:

- poor growth.
- weight loss.
- constant coughing and gagging.
- anemia.
- lowered production and reproduction.
- diarrhea or bloody feces.
- 7, worms in the feces.
- swelling under the neck.
- poor stamina.
- loss of hair and mange, the presence of a crusty skin condition caused by mites.
- visibility of the parasite itself.

Common Parasites of Small Animals

Parasites in general are very bothersome to both pets and the people that care for them. There are two types of parasites, internal parasites and external parasites. Internal parasites live inside animals, in the intestines, blood, muscle tissue or even brain. External parasites live on or in the skin. Common small animal parasites include heartworms, intestinal worms, fleas, ticks, mites and lice. All of these can cause a great deal of discomfort, illness and sometimes death.

Heartworms are internal parasites; the adult worms live in the heart and the progeny live in the bloodstream. The parasites are transmitted to your pet by the bite of a mosquito. The mosquito carries the larvae. The larvae eventually grow in the dog and becomes adult worms that lodge in the heart. The adult worms lodged in the vessels of the heart cause many problems for your dog or cat including coughing, vomiting, lethargy and weight loss. Eventually heart failure and death can occur. Treatment for the adult heartworms is available for dogs, but it comes with significant risks and is costly. Currently there is no safe treatment available for cats.

Fortunately, there are excellent preventative medications to stop this parasite. There is a monthly tablet, or chew treat which destroys the larva transmitted from the mosquito. We recommend that your dog be tested for heartworm in March or April to make sure your dog does not have the disease first. Giving the preventative to an infected dog may cause an allergic reaction.

Intestinal Parasites

There are many intestinal parasites. These parasites include roundworms, hookworms, tapeworms, whipworms, coccidian, giardia, and cryptosporidium. All can cause a large amount of damage to the intestinal lining leading to vomiting, diarrhea, anemia and weight loss. In puppies and kittens, death may occur if there is an enormous amount of parasites because the irritation interferes with proper nutritional absorption.

The transmission of these parasites varies. Pets may be infected through the placenta before the puppies or kittens are born, through the mom's milk during nursing, or from directly ingesting the egg of the parasite. Treatment depends on the type of intestinal parasitic infection. It is important to see a veterinarian if you're pet has any of the above symptoms. Often the intestinal parasites shed eggs in your pet's stool so veterinarians will check their stool to look for the eggs.

External Parasites

Fleas, ticks and mites are considered external parasites since they are found in or on the skin and hair. Fleas cause problems because they multiply very rapidly. Fleas are transmitted from pet to pe and wild animals to pet. The adult fleas take small bites of your pet and feed off the blood. This is not a large problem if there are just a few fleas but can lead to anemia and weight loss in young animals. Also, some pets are allergic to flea bites causing them to scratch until they have a skin infection.

There are many treatments for fleas. Veterinarians recommend the less toxic sprays, top spots and oral pills verus the powders and collars. It is also very important to clean and treat the environment and the pet's home to eliminate flea eggs, larva, and pupa at the same time you are treating your pet.

Ticks are concerning since they carry disease and transmit them to your pet. Some ticks carry Lyme disease and/or other infectious agents that cause diseases like Babesiosis, Rocky Mountain spotted fever; and tick paralysis. Lyme disease in dogs causes fevers, swollen lymph nodes, swollen joints, limping, lethargy and decreased appetite. The treatment for Lyme disease is a long course of antibiotics, usually 4-6 weeks. There is vaccine for Lyme disease that may help to prevent it, but its potency is not as good when compared with other vaccines. Available tick prevention includes sprays, top spots and collars. Preventing the ticks from biting, can stop the diseases by preventing the transfer.

Lice and mites live in and on the skin and hair follicles. Lice are usually considered more of a nuisance by causing hair and skin irritations. They are treated with shampoos, sprays and dips. Mites are generally species specific and are transmitted from one animal to another by contact. Some mites cause mange or scabies, which results in extensive scratching, hair loss, and inflammation of the skin. This scratching and inflammation leads to scabs and severe skin infections. Treatment involves injections of an insecticide or insecticidal dips and antibiotic for secondary bacterial infections.

PREVENTING AND TREATING ANIMAL HEALTH PROBLEMS

There are a number of activities and procedure that are used to prevent and treat health problems found in animals. Some of these include administering drugs, dipping, and restraining animals. The roles of feed additives and vaccination will also be explored. Vaccination is the injection of an agent into an animal to prevent disease.

Administering Drugs

There are several factors to be considered before administering drugs to an animal. These include determination of the amount to be administered, type of drug to use, purpose of the drug, site of administration of the drug, and type of animal to be treated. Most of this information can be found on the drug container. It is important that the drug manufacturer's recommendations be followed closely. Another factor that needs to be considered is the amount of time that the drug remains in the animal. This is important when determining how long milk from the animal will be contaminated by the drug. Contaminated milk must be discarded. Also, it must be determined how long to wait before a treated animal can be slaughtered for meat.

Drugs may be manufactured and sold as pills, powder, paste, or liquid (Figure .6).

Pills

The procedure for giving a pill to an animal is to restrain the animal and lift its head so that the mouth opens. Force the pill as far down the side of the mouth as possible, using either your hand or a balling gun (a device used to place a pill in an animal's throat). Massage the animal's throat until it swallows the pill.

Powders

These drugs are normally mixed in the feed or water of the animal. Often it is necessary to withhold feed or water for a period of time before administering the drug. Otherwise, the animal may refuse to eat or drink the drugged food.

Paste

Paste is normally used for treating horses for worms. The preparation is placed on the back of the horse's tongue with a caulking gun, and the horse is forced to swallow. Pastes are used for horses because it is often nearly impossible to treat them for worms by any other method.

Liquids

Liquid drugs administered orally (by mouth) are often placed directly in the animal's stomach by drenching. Drenching is the process of administering fairly large amounts of liquid to an animal. A syringe or drenching gun is used. In the process, the animal is restrained, with the head held level. The upper lip of the animal is lifted, and the tube is inserted along the side of the tongue. The drug is released, and the animal is allowed to swallow. Care must be taken not to get the drug into the animal's lungs.

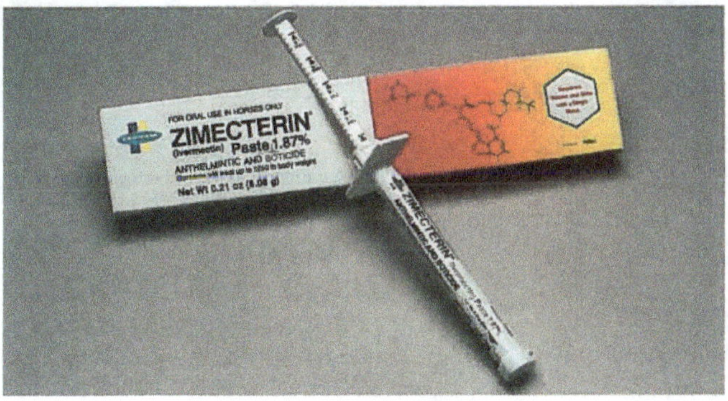

Figure .6: Pills. Liquids, powders, and pastes are methods of getting medicine to animals. It is important that the drug manufacturer's recommendations be followed closely.

The injection of drugs into animals takes many forms, based on the location of the injection. Injection is the process of administering drugs by needle and syringe. Some of the injection sites include intravenous (in a vein), intramuscular (in a muscle), subcutaneous (under the skin), intradermal (between layers of skin), intraruminal (in the rumen), and intraperitoneal (in the abdominal cavity). One determining factor as to where injections are made is how fast the drug needs to work. A drug injected into the blood is available faster than one injected under the skin. Often it is desirable for drugs to be released slowly over a long period of time. Growth hormones are generally administered in this way.

The procedure for giving an injection is to:

- Restrain the animal.
- Select the location for the injection.
- Fill the syringe, making sure that all air is removed.
- Disinfect the area to be injected.
- If the injection is to be made intradermally, clip the hair from
- The area to be injected.
- Insert the needle in the desired area without the syringe attached. (this prevents the loss of the drug if the animal jumps.)
- Attach the syringe to the needle and inject the liquid.

Infusion

Infusion is another method of getting drugs to the site of the infection. It is used most often to treat dairy animals with udder and teat problems. The udder is the milk-secreting gland of the animal. Teats are the appendages of an udder. A sterile cannula (blunt needle) is inserted into the opening of the teat, and the drug is forced into the teat canal (Figure .7).

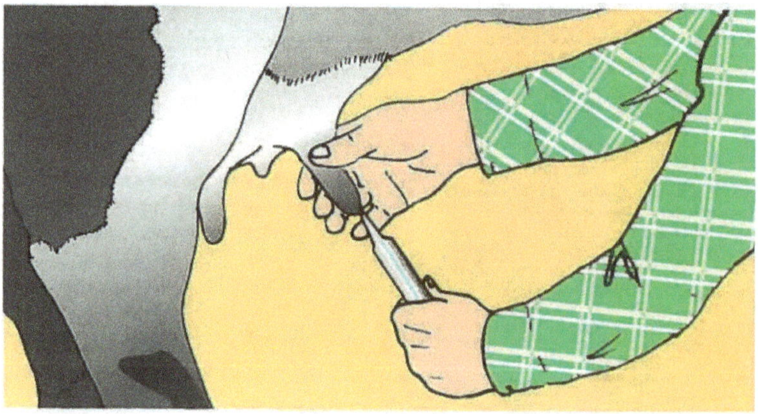

Figure .7: Infusion is a method of treating udder problems.

Dipping

Dippingis a process for treating animals, mostly cattle and sheep, for external parasites. It involves filling a vat with medicated water and forcing the animal to walk or swim through it. This process is also used to treat dogs for ticks and fleas. Dipping is popular where large numbers of animals must be completely covered with the medication.

Taking Temperatures

Taking an animal's temperature is basically the same as taking a human‘s temperature. It is usually taken in the rectum. The rectum is the last organ in the digestive tract. Animal thermometers are normally longer and heavier than those used in human medicine.

To use an animal thermometer, first shake down the column of mercury. Coat the thermometer with sterile jelly to make insertion easier. Do not force the thermometer into the rectum. If there is resistance, injury may result. Correct the conditions that are causing the resistance and then reinsert the thermometer. After several minutes, remove the thermometer and read the temperature on the scale. Thermometers with digital readouts are also available.

Determining Pulse and Respiration Rates

The pulse rate for a large animal can be taken by holding your ear against the animal's chest and listening to the heartbeat. The number of heartbeats in one minute is the pulse rate.

The respiration rate of an animal can be determined by watching its rib cage move. Counting the number of breaths that the animal takes in one minute will indicate the rate of respiration.

Restraining Animals

There are a number of ways to restrain animals for observation and treatment of diseases and parasites. These include head gates, squeeze chutes, halters, twitches, nose leads, and casting harnesses. Head gates trap the heads of large animals, whereas squeeze chutes hold the whole animal. When halters are used, they are usually tied to a post or something else substantial to hold the animal. Twitches hold the tender lip of a horse. Nose leads hold cattle by the nose (Figure .8). Sometimes, large animals must be lying down in order to be examined. An easy way to accomplish this is by using a casting harness (Figure .9). Properly applied, a casting harness will cause an animal to fall down with just a gentle tug of the rope.

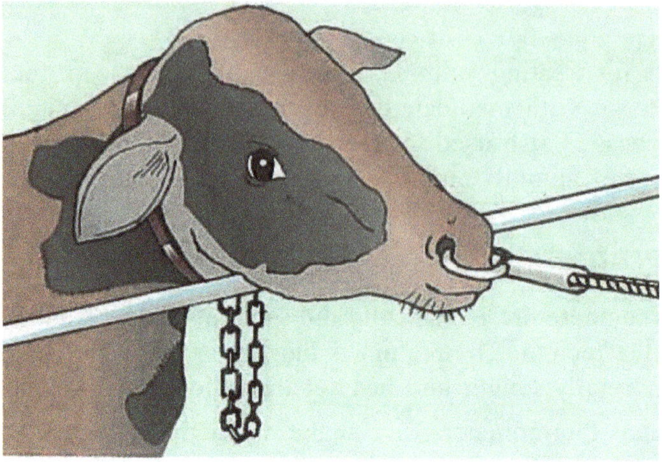

Figure .8: A nose lead used to restrain catle.

Figure .9: A properly used casting harness makes putting a large animal down a simple task.

Vaccination

The prevention of diseases is nearly always less expensive than treating animals once they have diseases. A good disease-prevention pro- gram should include vaccination of animals. Vaccination is the injection of an agent into an animal to prevent disease. The agent causes the animal's body to become immune to the disease. Immune means not affected by something. Vaccination programs are usually part of the services of a veterinarian (an animal doctor). Vaccination programs vary widely with the type of animal and area of the country.

Feed Additives

These additives are used primarily to control the incidence of low-level infections in growing animals. The materials used are primarily antibiotics that help increase feed efficiency and rate of gain, as well as control disease. Feed additives are sometimes used to control internal parasites. Caution must be taken to always follow the manufacturer's recommendations concerning the use of these materials. Failure to do so may lead to contamination of animal products used for human consumption.

REFERENCES

1. https://www.nidirect.gov.uk/articles/welfare-primates-physical-health
2. http://princetonvet.net/5-signs-your-farm-animals-may-be-sick/
3. http://media-out.vcpusd.net/9-12Resources/Ag1/28.%20%20Unit%2027,%20 Animal%20Health.pdf
4. http://www.merckvetmanual.com/mvm/management_and_nutrition/health-management_interaction_dairy_cattle/the_health_management_program_in_dairy_ cattle.html
5. http://www.fao.org/docrep/t0690e/t0690e04.htm#unit 5: appearance of the healthy animal
6. https://www.britannica.com/science/animal-disease
7. Angniman, P.A. 1996. Privatization of veterinary services within the context of structural adjustment in Mali, Cameroon and Chad. Rome, FAO.
8. Ashley, S.D., Holden, S.J. & Bazeley, P.B.S. 1996. The changing role of veterinary services: a report of a survey of chief veterinary officers' opinions. Somerset, UK, Livestock in Development.
9. Babjee, A.M. 1996. Privatization with reference to the veterinary services in Malaysia. Paper presented at the FAO/GTZ Workshop on Systematic Improvement of the Efficiency of Public and Private Livestock Services in Asia, 22-26 April 1996, Bangkok, Thailand.
10. Baker, J.L. 1995. Profile of veterinary services in New Zealand. Rome, FAO.
11. Cheneau, Y. 1984. Towards new structures for the development of animal husbandry in Africa south of the Sahara. Rev. Sci. Tech. Off. Int. Epiz., 3(3): 621-627.
12. Cheneau, Y. 1985. The organization of veterinary services in Africa. Rev. Sci. Tech. Off. Int. Epiz., 5(1): 107-154.
13. CTA (Technical Centre for Agriculture and Rural Cooperation). 1985a. Seminar on a Primary Animal Health Care Structure. 24-26 October 1984, Bujumbura, Burundi.
14. CTA (Technical Centre for Agriculture and Rural Cooperation). 1985b. Proceedings of the Seminar on Primary Animal Health Care in Africa. 25-28 September 1985,

Balantyre, Malawi.

15. de Haan, C. & Bekure, S. 1991. Animal health services in sub-Saharan Africa: initial experiences with alternative approaches. World Bank Technical Papers No. 134. Washington, DC, World Bank.
16. FAO. 1991. Guidelines for strengthening animal health services in developing countries. Rome.
17. FAO. 1994. Expert Consultation on Application of effective herd health and production programmes to increase livestock productivity in developing countries. 30 March-1 April 1993, Rome.
18. FAO. 1996. World livestock production systems - current status, issues and trends. FAO Animal Production and Health Paper No. 127. Rome.
19. FAO/RLAC. 1994. Report of the round table on privatization of the veterinary services in Latin America and the Caribbean. 12-14 November 1991, Santiago, Chile.
20. FAO/RNEA. 1994. Study on privatization of veterinary services in the Near East region. Report of a meeting, 28-30 June 1994, Cairo, Egypt.
21. Fassi-Fehri, B.E.C. & Bakkoury, M. 1995. Profile of the veterinary services in Morocco. Rome, FAO.
22. Gatongi, P.M., Scott, M.E., Ranjan, S., Gathuma, J.M., Munyna, W.K., Cheruiyot, H. & Prichard, R.K. 1993. Effects of three nematode anthelmintic treatment regimes on flock performance of sheep and goats under extensive management in semi-arid Kenya, Vet. Parasitology, 68: 323-336.
23. Holden, S., Ashley, S. & Bazeley, P. 1996. Improving the delivery of animal health services in developing countries: a literature review. Somerset, UK, Livestock in Development.
24. Howard, J.R. 1969. Costs of feedlot disease and prevention programs. J. Am. Vet. Med. Assoc., 154(10): 1166-1167.
25. Ilemobade, A.A. 1996. Privatization of veterinary services within the context of structural adjustment in Zimbabwe, Namibia and Ghana. Rome, FAO.
26. Jemal, A. & Hugh-Jones, M.E. 1995. Association of tsetse control with health and productivity of cattle in the Didessa Valley, western Ethiopia. Prev. Vet. Med., 22: 29-40.
27. Leonard, D.K. 1987. The supply of veterinary services: Kenyan lessons. Agric. Admin. & Extension, 26: 219-236.
28. Leonard, D.K. 1993. Structural reform of the veterinary profession in Africa and the new institutional economics. Dev. & Change, 24: 227-267.
29. McCorkle, C.M. & Mathias, E. 1996. Paraveterinary healthcare programmes: a global overview. In K.-H. Zessin, ed. Livestock production and diseases in the tropics: livestock production and human welfare. Proceedings of the VII International Conference of Institutions of Tropical Veterinary Medicine, Vol. II, p. 544-549. 25-29 September 1994, Berlin. Feldafing, Deusche Stifung fur Internationale Entwicklung.

30. McInerney, J.P., Howe, K.S. & Schepers, J.A. 1994. A framework for the economical analysis of disease in farm livestock. Prev. Vet. Med., 13: 137-154.

31. Mehraban, A.B., Barker, T. & Ward, D.E. 1996. Community-based animal health and production in Afghanistan - a delivery system. Paper presented at the Asian-Australian Animal Production Meeting, 13-18 October 1996, Makuhari, Japan.

32. Mukhebi, A.W., Kariuki, D.P., Mussukuya, E., Mullins, G., Nugumi, P.N., Thorpe, W. & Perry, B.D. 1995. Assessing the economic impact of immunisation against East Coast fever: a case study in Coast Province. Kenya. Vet. Rec., 137: 17-24.

33. OIE (International Office of Epizootics). 1995. Recommendation of the OIE: appropriate privatisation of veterinary practice in Africa. Rabat, Morocco.

34. Ott, S.L., Hillberg Seistzinger, A. & Hueston, W.D. 1995. Measuring the national economic benefits of reducing livestock mortality. Prev. Vet. Med., 24: 203-211.

35. Peagram, R.G., James, A.D., Bamhare, C., Dolan, T.T., Hove, T., Kan, G.K. & Latif, A.A. 1996. Effects of immunisation against Theileria parva on beef cattle productivity and economics of control options. Trop. Anim. Health Prod., 28: 99-11.

36. Schillhorn van Veen, T.W. & de Haan, C. 1995. Trends in the organization and financing of livestock and animal health services. Prev. Vet. Med., 25: 225-240.

37. Schreuder, B.E.C., Moll, H.A.J., Noorman, N., Halimi, C., Kroese, A.H. & Wassink, G. 1995. A benefit-cost analysis of veterinary intervention in Afghanistan based on a livestock mortality study. Prev. Vet. Med., 26: 303-314.

38. Sidahmed, A.E. 1995. Livestock development and rangelands management cluster: privatization of veterinary services in sub-Saharan Africa. Rome, International Fund for Agricultural Development (IFAD).

39. Swallow, B.M., Mulatu, W. & Leak, S.G.A. 1995. Potential demand for a mixed public-private animal health input: evaluation of a pour-on insecticide for controlling tsetse-transmitted trypanosomiasis in Ethiopia. Prev. Vet. Med., 24: 265-275.

40. Till, J.E. 1995. Building credibility in public studies. Am. Sci., 83(Sept-Oct): 468-473.

41. Umali, D., Feder, G. & de Haan, C. 1994. The balance between public and private sector activities in the delivery of livestock services. World Bank Discussion Papers No. 63. Washington, DC, World Bank.

42. Umali, D. & Schwartz, L. 1994. Public and private agricultural extension: beyond traditional frontiers. World Bank Discussion Papers No. 236, Washington, DC, World Bank.

43. Wamukoya, J.P.O., Gathuma, J.W. & Mutiga, E.R. 1995. Spontaneous private veterinary practices evolved in Kenya since 1988. Rome, FAO.

CHAPTER 4

Reproductive Anatomy and Physiology of Cattle

Chapter Contents

Optimal reproductive performance is fundamental to profitability of the dairy enterprise. Consequently, it is the focal point of dairy management and veterinary health programs on most successful dairy operations. When the average age of heifers at first calving is 24 versus 30 months, lifetime milk production potential is increased and replacement rearing costs decrease substantially. In cows, earlier conception following calving is rewarded by enabling cows to be dried off earlier. This results in fewer days dry and a higher average daily milk production. Efforts to attain these reproductive goals are consistent with progress toward the overall objective of achieving the maximum amount of milk yield per day of cow life.

INTRODUCTION

Successful artificial insemination programs are based on a clear understanding of the anatomy and physiology of reproduction in cattle. Before attempting to inseminate cows, you must develop a mental picture of the anatomical parts that comprise the female reproductive tract. In order to understand why an animal displays the many signs of estrus, when she should be inseminated, and how the pregnancy develops, you must clearly understand the hormonal mechanisms controlling the estrous cycle in cattle.

ANATOMY

First, let's look at the parts that make up the reproductive system in cattle (Figure). There are two ovaries, two oviducts, two uterine horns, a uterine body, cervix, vagina and vulva. The bladder lies below the reproductive tract and is connected at the urethral opening located on the vaginal floor. The rectum is located above the reproductive system. The vulva is the external opening to the reproductive system. The vulva has three main functions: the passage of urine, the opening for mating and serves as part of the birth canal. Included in this structure are the lips and clitoris. The vulva lips are located at the sides of the opening and appear wrinkled and dry when the cow is not in estrus. As the animal approaches estrus, the vulva will usually begin to swell and develop a moist red appearance. The vagina, about six inches in length, extends from the urethral opening to the cervix. During natural mating, semen is deposited in the anterior portion of the vagina. The vagina will also serve as part of the birth canal at the time of calving. The cervix is a thick walled organ forming a connection between the vagina and uterus (Figure). It is composed of dense connective tissue and muscle and will be the primary landmark when inseminating cattle. The opening into the cervix protrudes back into the vagina. This forms a 360° blind-ended pocket completely around the cervical opening. This pocket is referred to as the fornix. The interior of the cervix contains three to four annular rings or folds that facilitate the main function of the cervix, which is to protect the uterus from the external environment. The cervix opens anteriorly into the uterine body. About an inch long, the body of the uterus serves as a connection between the two uterine horns and the cervix. The uterine body is the site where semen should be deposited during artificial insemination. From the uterine body on, the reproductive tract separates and all further structures come in pairs (Figure).

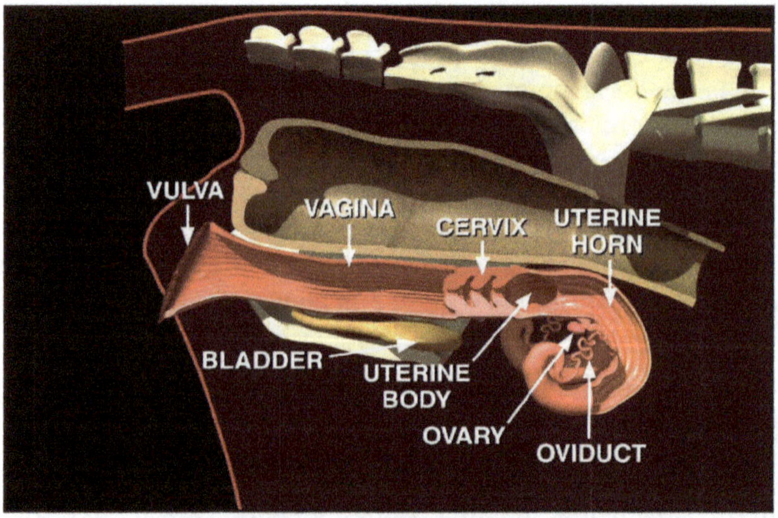

Side view of the cow's reproductive system

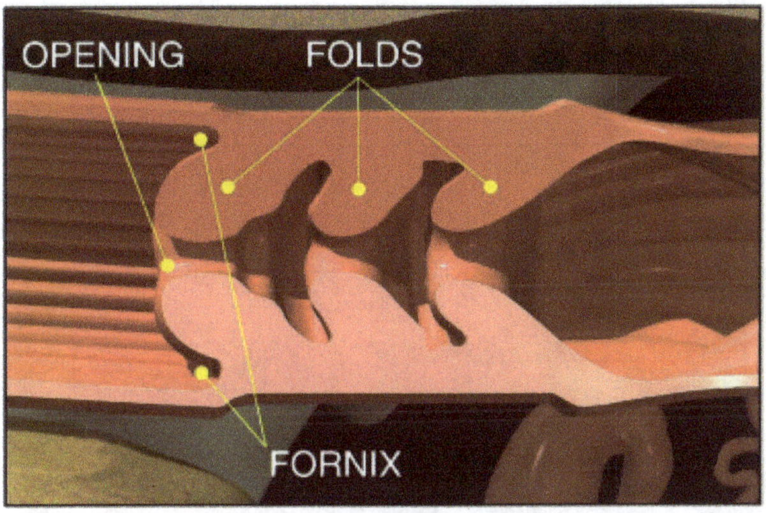

Close-up view of the cervix.

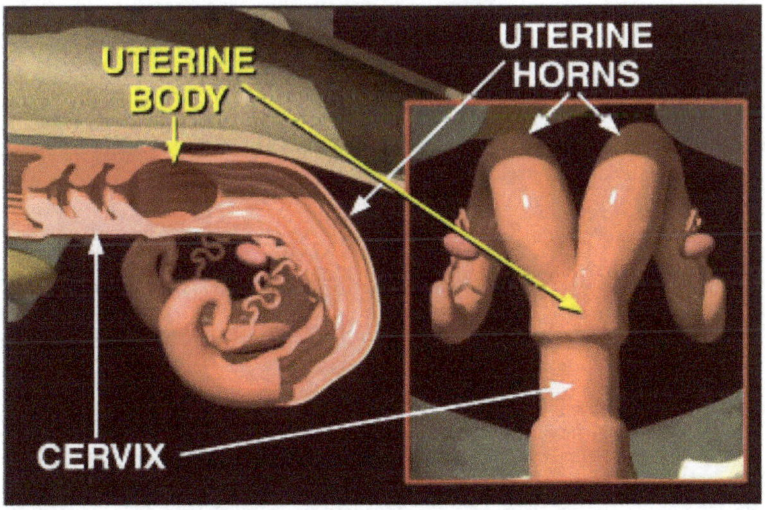

The short uterine body divides into two long uterine horns.

The two uterine horns consist of three layers of muscle and a heavy network of blood vessels. The main function of the uterus is to provide a suitable environment for fetal development. When a cow is bred, either naturally or by artificial insemination, the uterine muscles, under the influence of hormones oxytocin and estrogen, rhythmically contract to aid in sperm transport to the oviducts (Figure). Oviducts, as their name implies, carry ova, the cow's eggs. The oviducts are also commonly referred to as the fallopian tubes. The oviduct has several distinct regions when examined microscopically. The lower segment, closest to the uterus is called the isthmus. The connection between the uterus and the isthmus is called the utero-tubal junction or UTJ. The UTJ functions as a filter of abnormal sperm and the isthmus as a reservoir for healthy sperm (Figure). Research suggests that upon gaining access to the isthmus, healthy spermatozoa attach themselves to the walls.

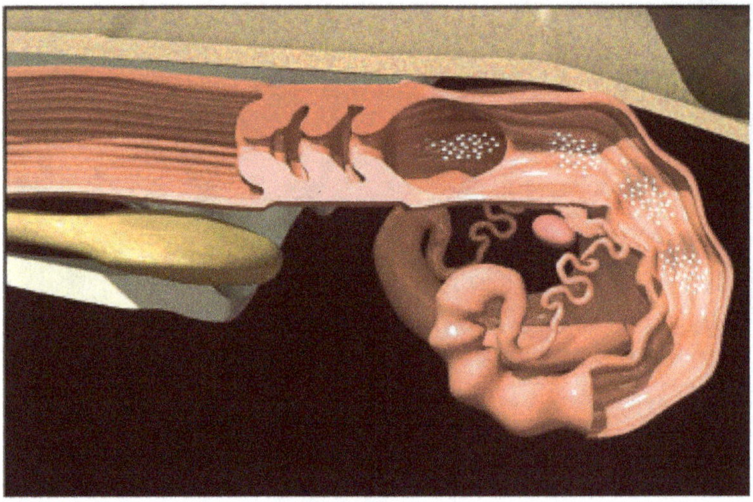

Uterine contractions aid in sperm transport

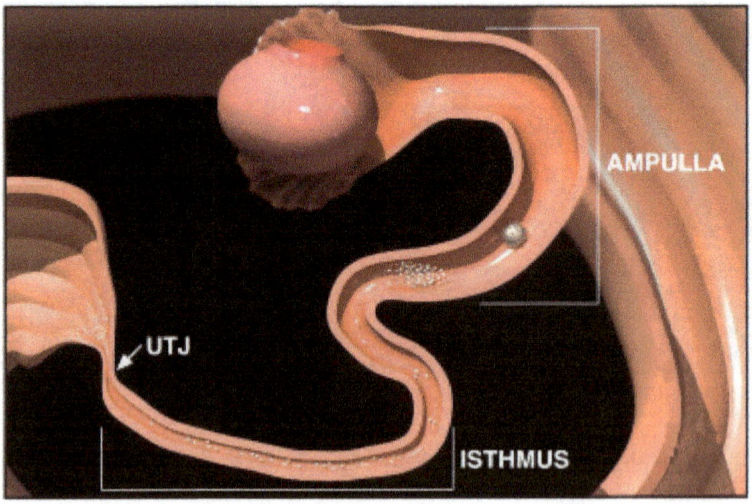

The UTJ, isthmus and ampulla are functionally different regions of the oviduct.

During this period of attachment, many physiological changes occur to sperm membranes, which are essential to attainment of fertilization potential. These changes are collectively referred to as capacitation and are apparently regulated by this very important attachment to the walls of the isthmus. It takes about five to six hours after insemination before a sufficient number of fertile sperm cells can populate the isthmus and complete the capacitation process. The upper portion of the oviduct, closest to the ovary, is referred to as the ampulla. The interior of the ampulla is more open than the isthmus allowing for easier passage of ova. It is within this segment of the oviduct that fertilization actually occurs. It is believed that a chemical signal released at the time of ovulation, stimulates the release of spermatozoa from the walls of the isthmus allowing them to continue their journey to the site of fertilization in the ampulla. The large funnel-like structure on the open end of the oviduct, called the infundibulum, surrounds the ovary, to recover the ova and keeps them from falling into the body cavity (Figure).

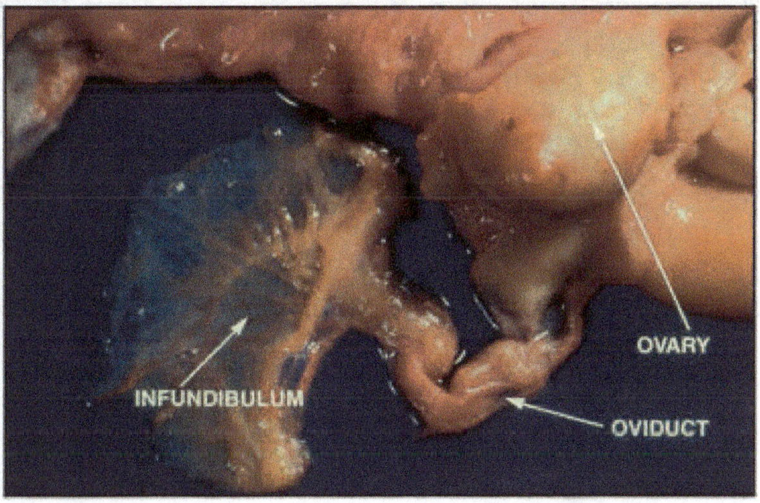

The infundibulum catches the egg from the ovary and guides it into the oviduct.

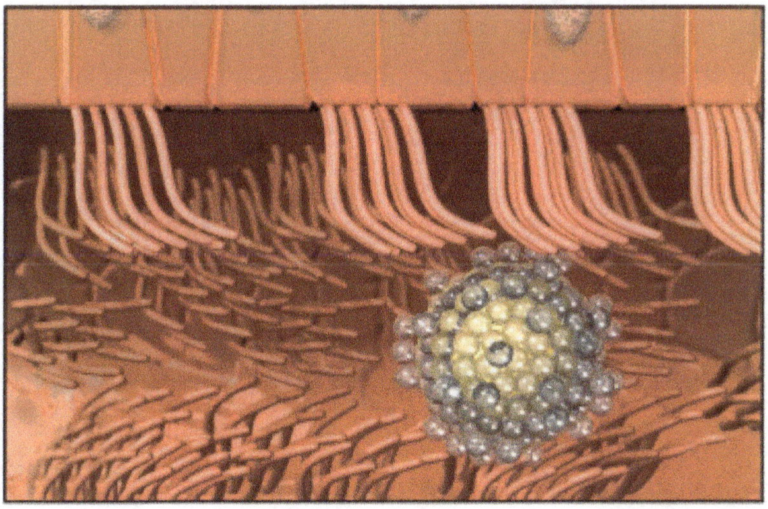

Egg and cumulus mass are transported into oviduct by hair-like structures

Hairlike structures on the infundibulum and within the ampulla rhythmically beat to move ova and a surrounding mass of cells called the cumulus mass down the oviduct to the site of fertilization (Figure). The ovaries are the primary organs in a cow's reproductive tract. They have two functions: to produce eggs and to produce hormones, namely estrogen and progesterone, throughout the different stages of the estrus cycle. On the surface of the ovary, you will usually find two different types of structures. Follicles are fluid filled, blister like structures that contain developing oocytes or eggs (Figure). Usually you will find numerous follicles on each ovary that vary in size from barely visible to ones 18 to 20mm in diameter. The largest follicle present on one of the ovaries is termed the "dominant follicle" and is the most likely candidate for ovulation when the animal comes into heat. Over time, greater than 95 percent of the other follicles on the ovary regress and die without ovulating and are replaced by new growing follicles.

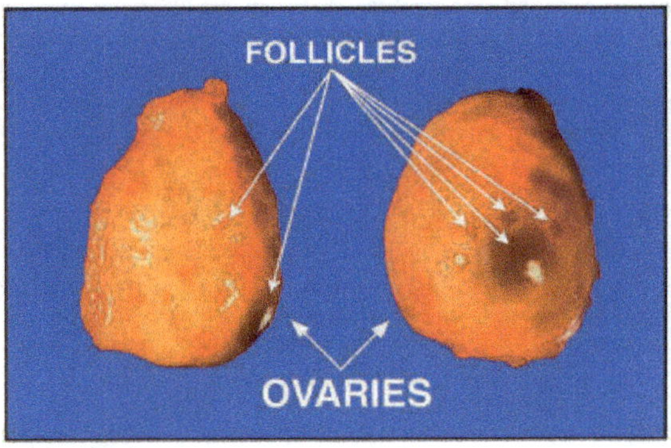

Blister like follicles on the ovaries

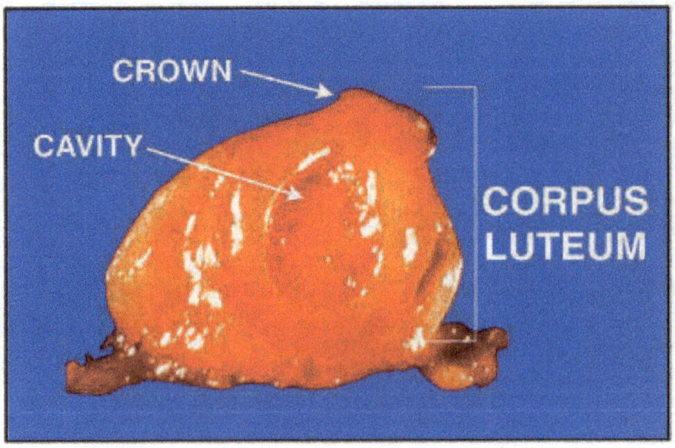

Cross section of bovine ovary with corpus luteum.

The other structure found on the ovarian surface is the corpus luteum or CL. The CL is the site where ovulation occurred during the previous cycle (Figure). Unless there were twin ovulations you should find only one CL located on one of the two ovaries. The CL will usually have a distinct crown protruding from the ovarian surface, which facilitates identification during rectal palpation. The CL may also have a fluid filled cavity but usually has a much thicker wall than a follicle and thus a much denser texture. "Corpus Luteum" is Latin for "yellow body." While the outside of this structure is usually dark red in appearance, a cross section reveals a bright yellow to yellow-orange interior.

THE ESTROUS CYCLE

Over a period of time, many changes take place in the reproductive system in response to changing hormone levels. These changes in normal open females repeat every 18 to 21 days. This regular repetitive cycle is called the estrous cycle (Figure). Let's discuss how the estrous cycle works starting with a cow in heat on day zero. Looking at the reproductive tract, we see several things happening. One ovary has a large follicle approximately 15 to 20mm in diameter.

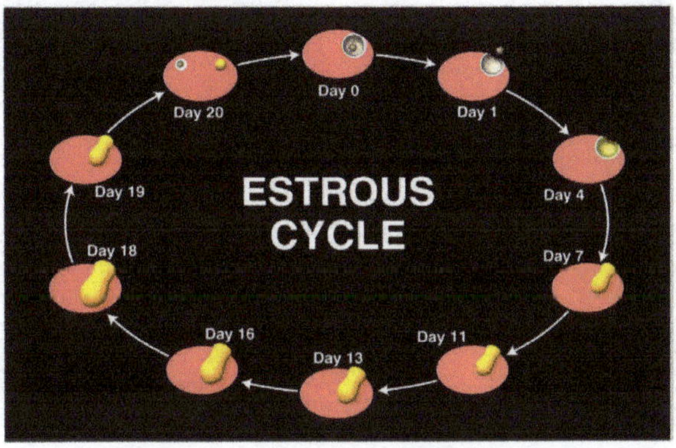

The bovine estrous cycle

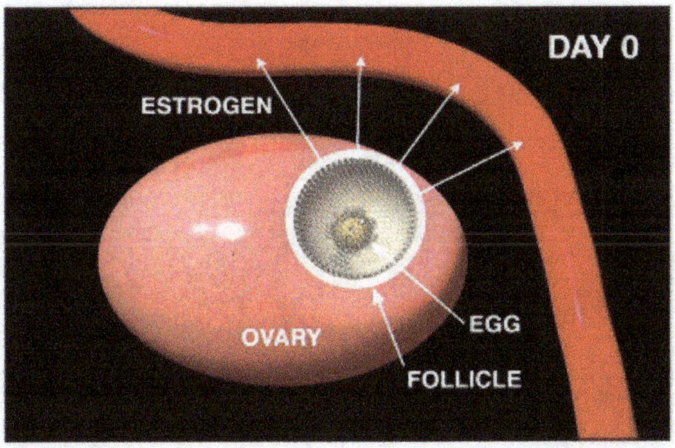

Estrogen from growing follicle is transported through the blood stream to all parts of the body.

This follicle has a mature egg inside ready to be released. The cells lining the follicle are producing the hormone estrogen (Figure 1). Estrogen is transported in the blood stream to all parts of the cow's body, causing other organs to react in a number of ways. It makes the uterus more sensitive to stimulation and aids in the transport of semen at the time of insemination. It causes the cervix to secrete viscous mucus that flows and lubricates the vagina. Estrogen is also responsible for all signs of heat including; a red swollen vulva, allowing other cows to mount her, going off feed, bellowing considerably and holding her ears erect are but a few of the many signs. On day one, the follicle ruptures or "ovulates" releasing the egg to the waiting infundibulum (Figure). Several hours prior to ovulation estrogen production declines. As a result, the cow no longer displays the familiar signs of heat. After ovulation, new types of cells called, luteal cells, grow in the void on the ovary where the follicle was located.

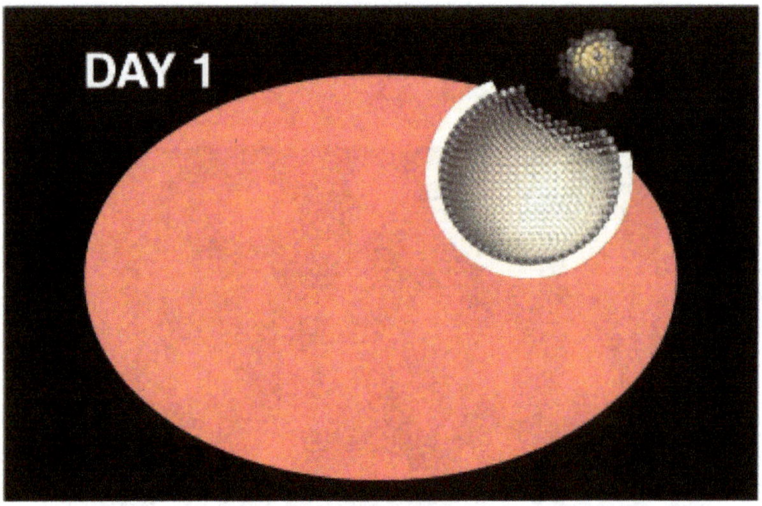

Luteal cells begin to grow in follicle void after ovulation.

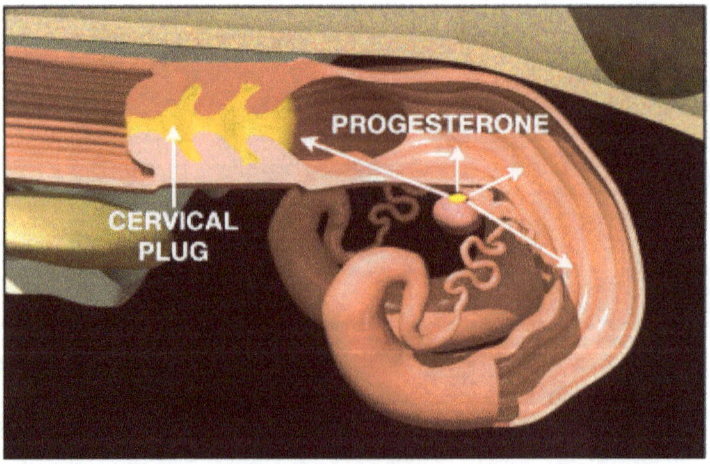

Progesterone from the corpus luteum prepares the uterus for pregnancy.

Quite rapidly over the next five to six days these cells grow to form the corpus luteum (CL). The CL produces another hormone, progesterone. Progesterone prepares the uterus for pregnancy (Figure). Under the influence of progesterone, the uterus produces a nourishing substance for the embryo called uterine milk. At the same time, progesterone causes a thick mucous plug to form in the cervix, preventing access of bacteria or viruses into the uterus. Progesterone also prevents the animal from returning to estrus by regulating the release of gonadotropins from the pituitary gland in the brain (Figure). There are two very important gonadotropins produced, stored and released from the pituitary gland. The first is follicle stimulating hormone, or FSH. As its name implies FSH stimulates the growth of small follicles. Luteinizing hormone or LH is the other gonadotropin. In addition to supporting progesterone production by the CL, LH can also stimulate estrogen production in large folicles. High levels of estrogen would bring the animal back into heat and make life difficult for a new embryo if she were pregnant.

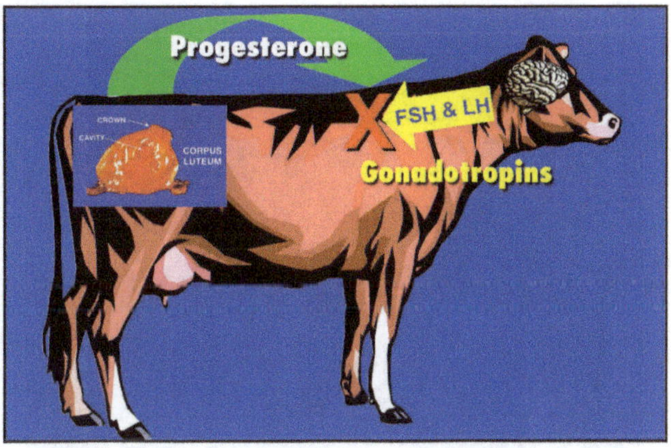

Progesterone regulates release of FSH and LH.

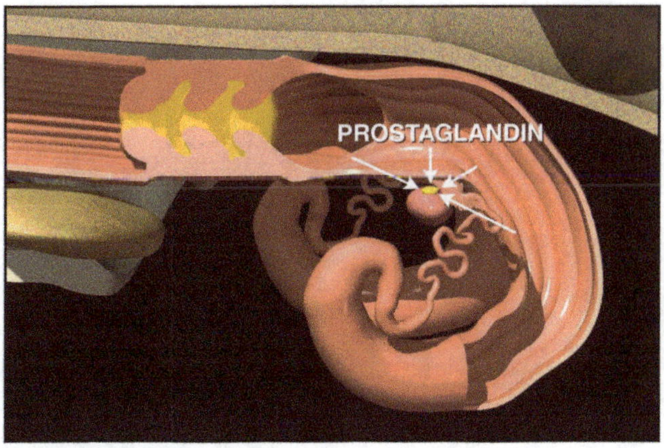

Prostaglandin destroys the corpus luteum.

Thus, progesterone's regulation of FSH and LH is a very important aspect of maintaining the pregnancy. On the other hand, if we did not breed the cow, we do want her to come into heat again. Days 16 through 18 of the estrous cycle are referred to as "the period of maternal recognition." During this time, the uterus searches itself for the presence of a growing embryo. If no embryo is detected, the uterus begins to produce another hormone called prostaglandin. Prostaglandin begins to destroy the CL (Figure). When the CL is destroyed, no more progesterone is produced and the pituitary gland begins to increase secretion of gonadotropins. Increased secretion of LH stimulates the dominant follicle to produce estrogen and bring the animal back into estrus (Figure . A full cycle is now completed. The average total time is about 21 days. The estrous cycle is subdivided into two phases based on the dominant hormone or ovarian structure during each phase. The luteal phase begins when the corpus luteum is formed; about five to six days after the cow was in heat, and ends when the CL regresses, about day 17 to 19 of the cycle.

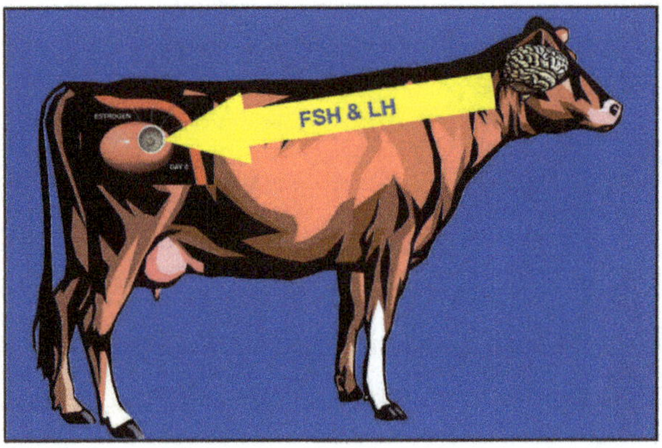

FSH stimulates growth of small follicles while LH stimulates proges- terone production by the CL and estrogen production by large follicles.

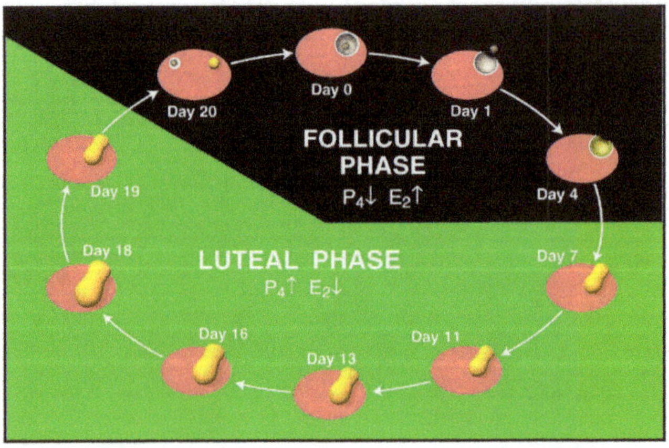

Estrous cycle is divided into two phases.

Progesterone levels are high during this phase of the cycle and estrogen levels are low. The other phase of the cycle; the follicular phase, begins when the CL of one cycle is regressed and ends when the new CL of the following cycle is formed. Thus, the follicular phase encompasses the period of time surrounding estrus (Figure). During this phase of the cycle estrogen levels are typically high while progesterone levels are low. As mentioned earlier, follicles may be present on the ovaries throughout the estrous cycle. Research using ultrasound technology has characterized follicular growth as occurring in "waves." Normally, an animal will have two or three waves of follicular growth during a 21-day cycle (Figure). The beginning of each wave is characterized by a small rise in FSH followed by rapid growth of numerous follicles. From this wave of follicles, one follicle is selected to grow to a much larger size than the others. This "dominant" follicle has the ability to regulate or restrict growth of all other follicles on the ovary.

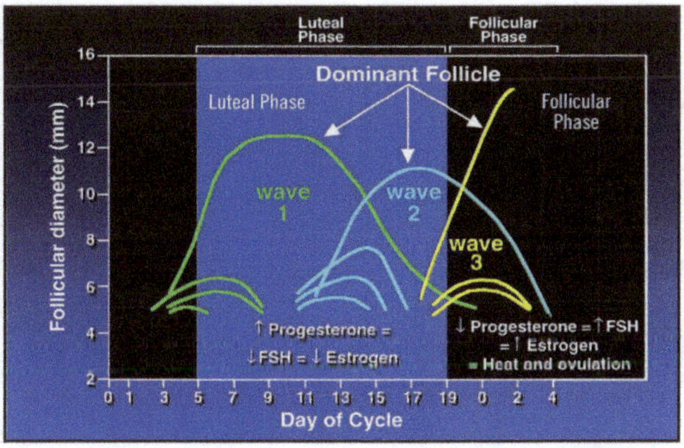

Follicular growth occurs throughout the estrous cycle but estrogen levels only rise during the follicular phase.

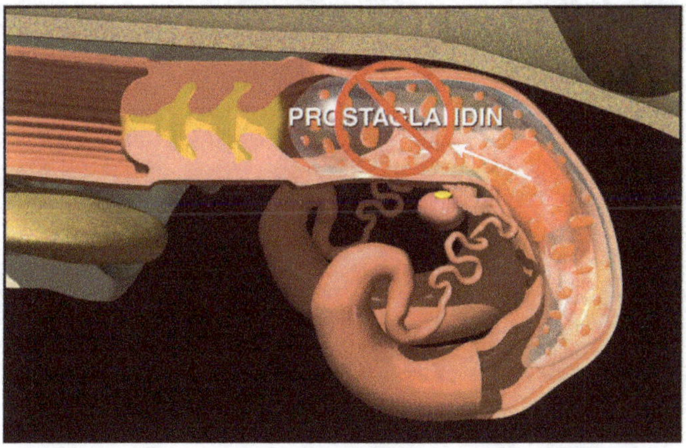

Fetus interferes with prostaglandin.

Dominant follicles only remain dominant for a short period of time, three to six days, which is followed by either cell death and regression or ovulation and release of the egg. Consequently, disappearance of the dominant follicle coincides with recruitment of the next wave of follicular growth. From this new wave another dominant follicle will be selected. Although it is typical for much follicular growth to occur throughout the estrous cycle, low levels of LH during the luteal phase, prevents these follicles from producing high levels of estrogen which would bring the animal back into heat. It is only the dominant follicle present at the time the CL regresses, when progesterone levels are low, that is permitted to produce enough estrogen to bring the animal back into estrus and to continue on to ovulation.

EMBRYO AND FETAL DEVELOPMENT

For the first four to five days, the embryo moves in the oviduct toward the uterus. Once the embryo gets there, it will be bathed in uterine fluids and continue to grow. While floating free in the uterus, several membranes, including the amnion, the chorion and the allantois are produced by the early embryo. Collectively, these membranes are referred to as the placenta. Hopefully, by the time the period of maternal recognition arrives, days 16 through 18, the fetus and growing placenta will have produced adequate quantities of the chemical signal required to maintain pregnancy. This signal interferes with the action of prostaglandin on the corpus luteum (Figure). The CL is thus retained and continues to produce progesterone, which is essential to maintenance of the pregnancy. At about 30 days of gestation, the placenta begins to attach to the uterus at several points. The placental sides of these attachment points are called cotyledons while the uterine side has caruncles. The attachment of cotyledons to caruncles is similar to Velcro.

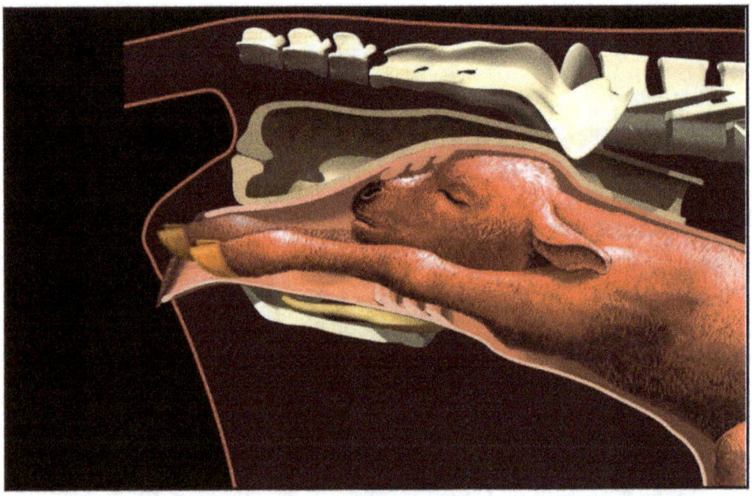

The end result.

This greatly increases the surface area within the attachment point, facilitating exchange of nutrients and waste products between calf and mother by way of arteries and veins leading to and through the umbilical cord. At calving, the muscles in the uterus begin to contract and eventually expel the calf and membranes through a dilated cervix and vagina. Several hormones including progesterone, estrogen, prolactin, relaxin and corticoids produced by the mother, the fetus, and the placenta, interact to bring about this event (Figure 0). Calving in a clean environment and proper treatment of the cow after a difficult calving will help prevent reproductive problems. Becoming familiar with the reproductive anatomy and physiology of a cow will help you do a better job artificially inseminating your cattle. Understanding how the hormones involved in the estrous cycle interact to control this phenomenon gives you clearer understanding why and when animals display the many signs of estrus, how pregnancy is maintained, and what you or your veterinarian must do to treat cows that do not cycle normally.

REFERENCES

1. ABELL - Annual Bibliography of English Language and Literature
2. ABES - Annotated Bibliography for English Studies
3. Bath Information Data Service
4. Bibliografia Nacional Espanola desde 1976 en CD-ROM
5. Bibliographie Nationale Française depuis 1970 sur CD-ROM
6. Bibliotecas sin Fronteras
7. Bodleian Library Pre-1920 Catalogue
8. British Library General Catalogue of Printed Books
9. Brotherton Collection Manuscript Verse
10. COPAC
11. Database of Classical Bibliography
12. Deutsche Nationalbibliographie
13. DocThèses
14. FABLE - Electronic Folklore Article Bibliography
15. Illustrated Incunabula
16. In Principio
17. International Medieval Bibliography on CD-ROM
18. ISSN Compact
19. Iter Italicum
20. Myriade - Le Catalogue Collectif National des Publications...
21. Novum Regestrum
22. Periodicals Contents Index
23. World Shakespeare Bibliography

CHAPTER 5

Breeding and Breeding Techniques

Chapter Contents

- **Introduction**
- **The History of Animal Breeding**
- **Animal Breeding Industries**
- **Method of Animal Breeding**
- **Selection by Nature**
- **Mating Systems**
- **Application of Crossbreeding**
- **Breeds of Cattle and Buffallo**
- **Selection Methods of Dairy Animals**
- **Herd Recording**
- **Breeding Strategies**
- **Crossbreeding Systems for Beef Cattle**
- **Artificial Insemination (Ai) Cows and Heifers**

Animal breeding, genetics, and genomics is the branch of science concerned with maximizing desirable genetic traits, such as producing animals that have leaner meat. Animal geneticists have identified elements within genes that can enhance animal growth, health, and ability to utilize nutrients. These genetic advances can increase production while reducing environmental impacts. Animal breeding is the selective mating of animals to increase the possibility of obtaining desired traits in the offspring. It has been performed with most domesticated animals, especially cats and dogs, but its main use has been to breed better agricultural stock. The more modern techniques involve a wide variety of laboratory methods, including the modification of embryos, sex selection, and genetic engineering. These procedures are beginning to supplant traditional breeding methods, which focus on selectively combining and isolating livestock strains. In general, the most effective strategy for isolating traits is by selective inbreeding; but different strains are sometimes crossed to take advantage of hybrid vigor and to forestall the negative results of inbreeding, which include reduced fertility, low immunity, and the development of genetic abnormalities.

INTRODUCTION

Animals and livestock contribute 40% of the global value of agricultural output and contribute to the livelihoods and food security of almost a billion people worldwide. Advances in animal breeding, genetics, and genomics are facilitating a more efficient industry. For example, the number of cattle has decreased over the past decade, yet the total production of beef and milk has increased. This was largely possible because genetic advancements led better animal feed efficiency (link is external), which is critical to improving livestock production and lowering costs for producers.

Breeding of high productive farm animals, like cattle, pigs and poultry is in the hands of multinational companies which invest a lot of money in state of the art breeding programs. The breeding of sheep, goats, horses and companion animals, e.g. the dog is based on individual breeders collaborating in the setting of a herd book or a breeder's association. Animal breeding is aiming at the improvement of animals by changing their genetic abilities for important traits. These traits are determined by the requirements and wishes from the society which might change over time. Animal breeding is highly influenced by research and developments in population-, quantitative- and molecular genetics. Sometimes, unexpected negative effects of animal breeding are observed that require adequate corrections. A breeding program will be presented here as a circular activity. Each generation, the program starts with formulating the breeding goal and ends with a critical review of the results obtained in the next generation. The evaluation might lead to a reconsideration of the breeding goal for the next round of selection.

THE HISTORY OF ANIMAL BREEDING

The history of animal breeding started even before the domestication of animals was ever recorded. In some cases, domestication was accidental while there were records of intentional selection for more friendly and tractable animals. Dogs, for example, were domesticated 12,000 years ago while horses were domesticated for draft and transportation purposes.

Robert Blakewell (1725-1795), a British agriculturist, is considered the father of animal breeding. He developed the Shire breed of horses, old Longhorn cattle, and Leicester sheep. He believed that "like begets like" since superior parents were more likely to produce superior progeny than were inferior parents. He suggested therefore to "breed the best to best.

" Jay Laurence Lush (1896-1982), an American animal breeder from the Iowa State University, on the other hand, is considered the father of modern animal breeding. In the 1930s, Lush established the foundation of modern methods of estimating breeding values. He asserted that "like does not always begets like." He argued that the progeny of best parents may be inferior to some progeny of the worst parents, because chance plays a part in the success of a particular mating. He therefore suggested to "breed the best to best and hope for the best."

After Lush, Charles Roy Henderson (1911-1989), an American animal geneticist from Cornell University and Alan Robertson (1920-1989), a British geneticist from the University of Edinburgh, became famous for their computer evaluations.

In the Philippines, popular breeds of farm animals such as the Philamin cattle, Berkjala pigs, and Los Banos Cantonese chickens were developed at the University of the Philippines College of Agriculture (UPCA) before the Second World War through the leadership of Filipino animal breeder Bienvenido Maria Sioco Gonzalez (1893-1953). In particular, the Los Banos Cantonese chicken was an imported strain of chicken from Canton, China and was consequently developed as a dual-purpose breed of chicken raised under Los Banos conditions. In 1930, the Los Banos Cantonese was exported to Linnan University, Canton, China and the University of Illinois, USA, marking the first time that an improved breed of livestock crossed the Pacific Ocean from the Philippines (Gonzalez 1932).

There are 5 very important aspects that should be considered in animal breeding:

- Most importantly, obviously for selective breeding to be successful it is essential that the trait (e.g. running speed or milk production or coat color) under selection is heritable.
- That animals have different genetic backgrounds so that selection is possible.
- The direction of selection is defined by humans and they decide which animals are allowed to mate and produce members of the next generation.
- Success of animal breeding can be judged by looking at a shift in population average phenotype from one generation to the next. So animal breeding works at population level, not automatically at individual level.
- Success of animal breeding can be measured as the cumulative result of multiple generations of selection. Breeding decisions are made with the future in mind.

Heritable Traits

Being able to predict the success of animal breeding relies on one very important factor that we still need to discuss: why does performance in offspring resemble that of the parents? Selective breeding will only be successful in case the trait under selection is *heritable*. Because only a fraction of the animals is selected for breeding, so is allowed to produce offspring, and because the trait is heritable, the performance in the offspring will resemble that of the parents. Therefore only the best parents are used for breeding and the average of the next generation will be better than that of the current. A trait is heritable if the performance for that trait, at least in part, depends on the genetic make-up (DNA) of an animal. Differences in performance between animals can (partly) be explained by genetic differences between animals. Animal breeding relates to intentional selection by humans based on animal performance in a certain environment for predefined and heritable traits. In most practical animal breeding schemes selection will be on more than one trait simultaneously. The animals that are superior in this combination of traits will be selected as breeding animals. In general this combination of traits will consist of traits related to performance (e.g. milk production, number of eggs, growth, and sport performance), health, and reproduction. The theory behind selection for a combination of traits easily becomes very complicated.

ANIMAL BREEDING INDUSTRIES

In many developed countries, systems of genetic improvement in most species have evolved with time, resulting in major changes in the structure of the industry. Breeding work is now concentrated in the hands of a few large companies where selection has been the main route to genetic improvement although much has been achieved by breed substitution and crossbreeding systems. By comparison, selection of livestock and poultry in the Philippines has been very limited and less effective in a dispersed and unresponsive breeding industry and one where most animals are raised by smallholder farmers. Like other developing country in the tropics, animal breeding programs are generally constrained by both physical and social environments.

The major inhibitors of genetic improvement include ill-defined breeding objectives, small population sizes, scarce capital resources and testing facilities in the long term, limited and ineffective use of artificial breeding and animal performance recording and evaluation services, and difficulty in demonstrating an advantage in economic merit of locally improved livestock. There is also a common lack of and/or discontinuity of long-term programs and commitment. Ignorance, indolence, illiteracy, tradition, superstition, politics, and lack of capital are also recognized as the major reasons why a national breeding program has not been successful. Improvement of livestock in developed countries was traditionally the role of the farmer-breeder working within a pedigree or breed association. It has never been organized into national breeding units or institutes, as has been common in plant breeding. In most countries, however, there are still many farmer-breeders who select on the basis of farm or centralized tests, and in some countries, use the help of artificial insemination (Al) stations. The organization of animal breeding industries as determined by the number of breeders or breeding companies/establishments involved in the genetic improvement of livestock and poultry is inversely related to the female reproductive rate of a particular species (figure). In cattle and carabaos for example, almost every producer is also a breeder. On the other hand, genetic improvement of commercial chickens in the world is concentrated in the hands of less than twenty large breeding companies.

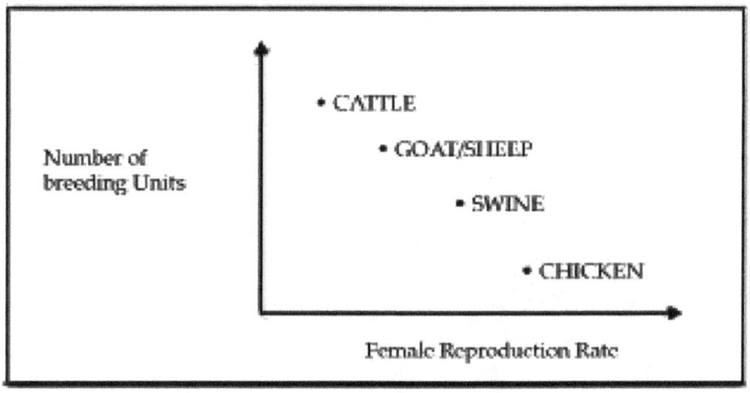

Figure : Relationship between female reproductive rate and number of breeding establishment for various livestock species.

Improvement programs for cattle and carabaos commonly associated with large numbers of breeders are usually run by national or regional/provincial organizations, most of which have a strong element of government intervention. Similar set-ups may also be noted for other livestock species owned largely by smallholder farmers. Genetic and breeding expertise is primarily made available at the headquarters of the organization. Such an organization often involves Al units that may employ highly trained professional breeders. Uniform recording and selection criteria are applied throughout the subject region, with emphasis on selection among bulls of known breeding value. Through Al, the reproductive rate of the bull is enhanced since one bull can be mated to several hundred or thousand females. In beef cattle, the farmer-breeder is still dominant. There is much performance testing (for growth rate) but little consistent selection. In dairy cattle, improvement is centered on progeny testing of bulls at AI stations, with farmer-breeders owning the females. Selection is accurate and effective on progeny-tested bulls, but selection of females is less satisfactory.

In contrast, breeding of chickens is now carried out almost entirely by a few breeding companies which compete internationally. Poultry breeding companies are usually privately owned, with primary breeding populations that run into hundreds or thousands of birds and multiplied into millions of commercial hybrid broilers and layers. They each maintain a number of stock under intensive testing and selection, and produce a range of commercial products for different production-marketing systems. Genetic improvement in the poultry breeding operations is specialized for the multiplier flock owners (multiplication), hatcheries (reproduction), and distributors (production). The breeding procedures are generally directed by highly-trained professional geneticists. It is a policy of most companies, however, not to disclose their precise breeding techniques. When increased female reproduction rates become possible in other farm species, a shift in the character of the corresponding breeding industries is anticipated towards today's situation in poultry. The swine breeding industry, with its intermediate reproductive rate, is beginning to resemble or is already similar to the poultry breeding industry.

METHOD OF ANIMAL BREEDING

Animal breeding is a branch of animal science that addresses the evaluation of the genetic value of domestic livestock. A breed is a group of domestic animals with a homogeneous appearance, behavior, and other characteristics that distinguish it from other animals.

Pure-Breeding

Pure-breeding is the mating of rams and ewes of the same breed or type. A purebred flock can be managed as a single flock because all ewes and rams are of the same breed. The goal of purebred sheep production is to provide superior genetics (seed stock) to the commercial sheep industry. Seed stock are marketed as rams and replacement ewes to other seed stock producers or to commercial sheep operations.

Improvements in purebred sheep should be documented through performance records. The National Sheep Improvement Program (NSIP) collects performance data from purebred producers and provides them with across-flock EPDs. "EPD" is short for "Expected Progeny Difference."

An EPD is an estimate of the genetic merit of an animal for a given trait. It is the expected difference between the performance of an animal's progeny (offspring) and the average progeny performance of all the animals in the breed for that trait. For example, if a lamb has a weaning weight EPD of 5.0 lbs., this means that its progeny will be 5.0 lbs. heavier than the average lamb in the breed.

Out-Breeding

Within pure-breeding, there are several types of mating systems. Out-breeding is the mating of animals of the same breed but which have no closer relationship than at least 4 to 6 generations. Outbreeding is the recommended breeding practice for most purebred sheep breeders.

Inbreeding

Inbreeding is a system of breeding in which closely related animals are mated. This includes sire to daughter, son to dam, and brother to sister. Technically, inbreeding is defined as the mating of animals more closely related than the average relationship within the breed or population concerned. The primary genetic consequence of inbreeding is to increase the frequency of pairing of similar genes.

Inbreeding is essential to the development of prepotent animals — animals that uniformly "stamp" their characteristics on their progeny. Inbreeding may also be used to uncover genes that produce abnormalities or death — genes that, in outbred herds, are generally present in low frequencies. Inbreeding is suggested for only highly qualified operators who are making an effort to stabilize important traits in a given set of animals.

In general, inbreeding results in an overall lowering in performance: vigor, disease resistance, reproductive efficiency, and survivability. It also increases the frequency of abnormalities. For example, the spread of spider lamb disease in black-faced sheep is believed to be the consequence of inbreeding.

Line breeding

Line breeding is a system of breeding in which the degree of relationship is less intense than in inbreeding and is usually directed towards keeping the offspring related to some highly prized ancestor. The degree of relationship is not closer than half-brother half-sister matings or cousin matings, etc. Line breeding is a mild form of inbreeding.

Crossbreeding

Crossbreeding is the mating of rams and ewes of different breed compositions or types. However, it does not denote indiscriminate mixing of breeds, but rather is a systematic utilization of different breed resources to produce crossbred progeny of a specific type. Crossbreeding is used extensively in the commercial sheep industry and the majority of slaughter lambs are crossbred.

Crossbreeding offers two distinct advantages:

- heterosis; and
- breed complementarity.

Heterosis

Heterosis, or "hybrid vigor," is defined as the superior performance of an offspring over the average of the parent breeds. This can have a marked effect on the profitability of a cattle operation. Heterosis is greatest when crossing two parent animals of totally unrelated ancestry. Hybrid vigor can be exhibited through a variety of traits such as increased survivability and growth of crossbred calves or higher reproduction rates of crossbred cows. The amount of heterosis that is maintained in a herd depends on the type of crossbreeding system the producer selects.

Breed Complementarity

The second major advantage of crossbreeding lies in the ability to utilize breed complementarity. All breeds have strengths and weaknesses. No one breed excels in all relevant traits. Thus, production can be optimized when mating systems place breeds in roles that maximize their strengths and minimize their weaknesses.

Mating Polypay ewes to Suffolk rams is an example of matching complementary strengths of breeds to optimize efficiency of a production system. This cross takes advantage of the reproductive efficiency and moderate maintenance costs of Polypay ewes while producing Suffolk-sired lambs to meet market requirements for fast-growing, heavy muscled lambs.

The efficiency of this cross would be much greater than the reciprocal mating of Suffolk ewes to Polypay rams. The latter cross would produce genetically equivalent market lambs (half Suffolk and half Polypay), but fewer lambs would be sold and production costs greatly increased due to higher feed requirements of heavy Suffolk ewes compared to Polypay ewes.

Key Issues in Animal Breeding

- In animal breeding people select the animals that will produce the next generation of animals that are on average superior than the present generation of animals.
- Selection by nature, natural selection, is very important for the adaptation of animals to the circumstances they are kept.
- An important prerequisite for the success of animal breeding is that traits are heritable, which implies that the ability for the trait is passed on from parents to progeny.
- Domestication of animals started with the dog. Later on the farm animals were domesticated. Domestication of a species requires specific traits. Within the domesticated species the process is still going on due to new requirements caused by changing circumstances.
- Selective animal breeding started 250 years ago with the formation of breeds and herd books. The scientific base for animal breeding is developed in the 20-th century. The application of scientific developments in reproduction in the past 50 years made animal breeding more effective. Recently it got a strong impulse from developments in molecular genetics.
- Breeding activities are directly be influenced by and are related to developments in society: required food production by animals and requirements of mankind for animals as companion and for leisure purposes.
- Breeding programs with cattle, pigs and poultry yielded sharp increases in the quantity of milk, meat or eggs. Also capabilities of horses improved sharply.
- Breeding of animals did not only give positive results: in dogs inbreeding and one-sided selection for conformation traits resulted in dogs with more health problems and less welfare. In farm animals selection for yield implies a risk for a deterioration of quality and fitness traits.

SELECTION BY NATURE

It sounds as if animal breeding is all in the hands of the humans. Compared to natural populations this indeed is the case, as we decide which animals are allowed offspring and which are not: selective breeding or in other words artificial selection. However, as in natural populations there is another force that plays an important role and that is the force of natural selection. In **natural selection** it is not us but the environment that determines survival and reproductive success of animals. So after we have decided which animals we intend as parents, they still need to be able to survive until reproductive age and to be able to reproduce successfully. As you can imagine, natural selection also results in directional change in the population average. Animals adapt to their environment and the ones who can do that best will be the most successful in survival and reproduction. In other words: in natural selection the direction of selection is on adaptation to the environment.

Even though animal breeding is defined as intentional selection by humans, you can see that natural selection will also play a role. In some cases natural selection will even work in the opposite direction of selective breeding. In those cases without human intervention the animals with the desired qualities will be less successful in surviving and/or producing offspring. For example, the fact that in many cows there is a negative relationship between high milk production and being able to get pregnant, shows that animals with the desired quality: high milk production, are less likely to produce offspring unless there is extra effort put into it by the farmer. Also, the fact that very high producing cows often have health problems indicates that their chance of producing offspring is reduced compared to their more average producing 'sisters'. Selective breeding often competes with natural selection. We have become so familiar to the fact that some of the best animals in a breed require assistance with some aspects of their survivability and/or reproduction that we think it is normal. Domestic animals are 'created' by humans and to maintain that we accept certain disadvantages.

Domestication and Animal Breeding

Domestic animals need to live in (close) association with humans, therefore they have to become tame. They also have to meet the expectations of their owners for the purpose of keeping them. This can be achieved by selective breeding. Expectations of owners will change in time, followed by a change in selective breeding plan. Domestication often has resulted in a type of animals that has become quite different from their wild counterparts.

As a result domestication often also involves the development of a dependency on humans so that the animals lose their ability to live in the wild.

Domestication of other Species

The symbiotic relationship, such as between human and the ancestors of the dog, most likely is quite unique for dogs. Other types of animals may have been domesticated more forcefully. They were captured and put in an enclosure or tied up, at least during the night, and were only allowed out to graze or scavenge under supervision of a herdsman. Only those animals that were not aggressive, but also not too shy, managed to adapt to these new circumstances. So (mainly natural) selective breeding also in these cases was on temperament. In the table 5.1 you will find a list of domestic animals with their approximate time and location of domestication. This is approximate, because especially for ancient times it is difficult to make an accurate estimate. But even recent events are not always straightforward. Because when do you call an animal domesticated? And what if it happened in more than one place simultaneously but independently of each other?

Table 5.1. List of domesticated animals in early times.

Species	Latin	Date	Location
Dog	*Canis lupus familiaris*	>30,000 BC	Eurasia
Sheep	*Ovis orientalis aries*	11000 - 9000 BC	Southwest Asia
Pig	*Sus scrofa domestica*	9000 BC	Near East, China, Germany
Goat	*Capra aegagrus hircus*	8000 BC	Iran
Taurine cattle	*Bos primigenius Taurus*	8000 BC	India, Middle East, North Africa
Zebu cattle	*Bos primigenius indicus*	8000 BC	India
Cat	*Felis catus*	7500 BC	Cyprus and Near East
Chicken	*Gallus gallus domesticus*	6000 BC	India and South East Asia
Llama	*Lama glama*	6000 BC	Peru
Guinea pig	*Cavia porcellus*	5000 BC	Peru
Donkey	*Equus africanus asinus*	5000 BC	Egypt

Domesticated duck	*Anas platyrhynchos domesticus*	4000 BC	China
Water buffalo	*Bubalus bubalis*	4000 BC	India, China
Horse	*Equus ferus caballus*	4000 BC	Eurasian Steppes
Dromedary	*Camelus dromedaries*	4000 BC	Arabia
Honey bee	*Apis*	4000 BC	Multiple places
Silkworm	*Bombyx mori*	3000 BC	China
Reindeer	*Rangifer tarandus*	3000 BC	Russia
Rock pigeon	*Columba livia*	3000 BC	Mediterranean Basin
Goose	*Anser anser domesticus*	3000 BC	Egypt
Bactrian Camel	*Camelus bactrianus*	2500 BC	Central Asia
Yak	*Bos grunniens*	2500 BC	Tibet
Asian elephant	*Elephas maximus*	2000 BC	Indus Valley civilisation
Alpaca	*Vicugna pacos*	1500 BC	Peru
Ferret	*Mustela putorius furo*	1500 BC	Europe
Common carp	*Cyprinus carpio*	Unknown	East Asia
Domesticated turkey	*Meleagris gallopavo*	500 BC	Mexico
Goldfish	*Carassius auratus auratus*	Unknown	China
European rabbit	*Oryctolagus cuniculus*	600	Europe
Japanese Quail	*Coturnix japonica*	1100–1900	Japan

Canary	*Serinus canaria domestica*	1600	Canary Islands, Europe
Fancy rat	*Rattus norvegicus*	1800s	United Kingdom
Fox	*Vulpes vulpes*	1800s	Europe
European Mink	*Mustela lutreola*	1800s	Europe
Cockatiel	*Nymphicus hollandicus*	1870s	Europe
Zebra Finch	*Taeniopygia guttata*	1900s	Australia
Hamster	*Mesocricetus auratus*	1930s	United States
Silver Fox	*Vulpes vulpes*	1950s	Soviet Union
Ball python	*Python regius*	1960s	Africa
Red Deer	*Cervus elaphus*	1970s	New Zealand
Atlantic Salmon	*Salmo salar*	1969	Norway
Atlantic Cod	*Gadus Morhua*	On going	Norway

MATING SYSTEMS

Mating systems are descriptions of who mates with whom in the animal world. In simplest terms, definitions of mating systems are based on how many mates an individual acquires during the breeding season. In monogamy, both males and females have only one mate at a time. This type of mating system often occurs in species in which both the male and female are required to successfully raise young or in which males have little chance of monopolizing more than one female. Monogamy is common in birds whose males can help incubate eggs and feed young.

We move on to consider the evolution of mate choice, and then we describe the types of mating systems found in animals.

The Evolution of Sex

Asexually reproducing animals pass on all of their chromosomes, and consequently all copies of each gene, to their offspring. In contrast, due to meiosis, diploid sexually reproducing animals have two copies of each chromosome but only pass one copy of each chromosome on to an egg or sperm cell. This means that a sexually reproducing diploid animal only passes half of its total genes on to its offspring. Despite the cost of losing half of the potential passage of genes to the next generation, sexual reproduction is much more common than asexual reproduction among animals because it provides several evolutionary advantages. The major advantage of sexual reproduction comes from genetic recombination. Genetic recombination allows an organism's offspring to be genetically diverse. Sexual reproduction increases the chances of acquiring favorable mutations and is unlikely to propagate deleterious ones. Genetic diversity within a group of offspring is advantageous as the local environment changes. This idea becomes clear when we examine organisms that can reproduce both sexually and asexually. Aphids, for example, will favor asexual reproduction when their environment is stable. When the environment is going to turn cold, most species of aphids reproduce sexually, because sexual reproduction produces eggs that are freeze tolerant and can diapause during the winter. Genetic diversity may also lead to evolved defenses against parasites and disease. The mud snail, Potamopyrgus antipodarum, is host to several trematode parasites. Sexual individuals of this species are more common in areas where risk of trematode infection is high. In areas where the risk of infection is low, asexual individuals have displaced sexual ones. This suggests that the genetic diversity acquired from sexual reproduction is necessary for this species to defend against parasites, as asexual individuals may not easily survive in areas where parasites are high.

Sexual reproduction often involves evolutionary differentiation of males and females. Females typically produce significantly fewer gametes (eggs) than males and invest heavily in each one. On the other hand, males produce many gametes (sperm) and invest little into each one. These strong differences in gamete investment between the sexes leads to reproductive strategies between the sexes that, in some cases, conflict. Females may spend more care than males selecting a mate due to the high cost of their gametes.

Variance in Mating Success and Bateman's Principle

A key element of the study of mating systems is understanding how many mates an animal has in its lifetime. Bateman's principle helps to make predictions about mating success and number of mates. Bateman's principle postulates that variance among females in mating success is low, whereas variance among males in mating success is high. This stems from the fact that one mating in females should be enough to fertilize all their eggs whereas in males reproductive success is based on the number of times they have mated. In other words, nearly all females in a population mate and have offspring, but relatively few males mate successfully (Figure). Those males that do mate tend to mate with many females-thus a few males have very high reproductive output, but many males have little or no reproductive output. This leads to the prediction that sexual selection should act more strongly on males, leading to greater elaboration of behavior and structures used in attracting mates in males than in females.

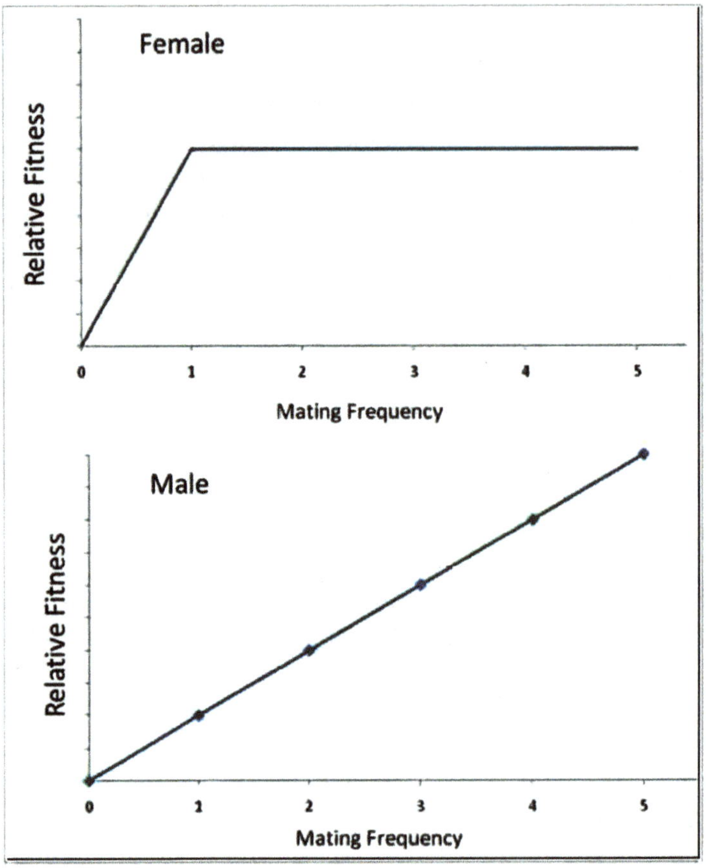

Figure : Bateman's principle.

Criticisms of Bateman's theory focus on the generality of the predictions. Contrary to the predictions of Bateman's principle, there are several possible advantages to female multiple matings. The female cichlid fish Pseudotropheus spiliopterus mates with any male they meet because they have a high risk of getting predated and a small population. This often leads to multiple matings by a single female. Mating with any male that is seen ensures that these cichlids have a chance at producing offspring. The female Malawi blue cichlid has a high population but still participates in multiple matings. In this case multiple matings occur to avoid inbreeding and increase genetic diversity among the offspring. Additionally, multiple matings by females may increase the likelihood that they will find a compatible mate, one that is not sterile, or even help prevent infanticide.

Female Mate Choice

Mate choice is also a key element of mating systems. In most species, females are choosier when picking a mate than males. A significant reason for this is the higher investment females make in each gamete than males. Females may prefer certain males for a variety of reasons, including "good genes", meaning that the male has attributes which predict better survivorship of the offspring, good potential parenting by the male, or possession of resources by the male that will support the offspring during their growth and development.

Additionally, in most species, females are more likely to provide parental care. Females that carefully select their mates are at a lower risk of losing their reproductive investment. Males may be under strong selection for certain traits that are favored by females. Most females look at these traits as indicators of their partner's fitness. Selection favors females that choose males that enhance the likelihood of her offspring's success. Males with more elaborate ornamentation, or that are more colorful, can be displaying a good indicator of value as a mate, and may win the chance to mate with a particular female. Although mating is important, it can be a costly event — females are predicted to be choosier about selecting their mates than males because of risks during mating, such as aggression or disease transmission, which may negatively impact the female's reproductive output.

Male Mate Choice

The importance of male mate choice is controversial. Older theory predicts that male mate choice should be less common in animals. However it plays an important role in many mating systems, and the cost of mating for males may have been underestimated in earlier studies. Male mate choice occurs most often when males are substantially involved in caring for their offspring, or when there is great variation in the quality of the females as mates within a population. If males are choosy about their mate, then over time females may evolve ornamentation or coloration that is subject to sexual selection.

Types of Mating Systems

There are four main mating systems:

Monogamy

Social monogamy is the behavioral pairing of a single male with a single female. It is most common in birds and rare in other animals. Theoretically, individuals in monogamous pairs will both contribute to the defense and parental care of offspring. Choosing an inappropriate mate could have a high fitness cost. Because the costs of poor mate choice in monogamous species can be so high, in some instances organisms engage in strategies of either serial monogamy or extra-pair copulations. Extra-pair copulations are very common in birds. Monogamy reduces the potential for genetic variation among a female's offspring. By mating with more than one male over the course of her lifetime, a female gains higher genetic variation among her offspring. The benefits of monogamy, which are shared parental care and territorial resources, are maintained by having only one mate at a time, or by concealing extra-pair partnerships.

Polygyny

Polygyny is the association of one male with multiple females. This mating system is found in a few birds and insects, but is most common in mammals. Polygyny is a strategy used by males to increase their reproductive fitness.

Resource Defense Polygyny

In resource defense polygyny, groups of females are attracted to a resource — males then compete for territorial possession of the resource, and, by extension, mating priority with females at the resource. Thus, individual males form territories centered on resources needed for successful mating.

Harems

Another common type of polygyny is membership in a harem, a defended group of females associated with one male. Females may initially associate in a harem for group defense, or they may be herded together by a male. Males compete for control of the groups. Harems typically exhibit a dominance hierarchy among the females in the group.

Leks

A **lek** is an aggregation of males that are each seeking to attract a mate. Within a lek, males typically perform sexual displays. Unlike most other mating systems, leks are not associated with resources. Aggregations of males may be near particularly attractive females or in areas where females are likely to travel. It is thought that males form leks because they attract more females than do isolated males. Attracting more females is a strategy used by males to help increase their reproductive success.

Polyandry

Polyandry is a group with one female and many males. Polyandry is a reproductive strategy that helps a female ensure reproductive success by providing her with multiple mating options.

Resource Defense Polyandry

In the Spotted Sandpiper, females control resources, which in turn controls male mating associations.

Cooperative Polyandry

The Galapagos hawk exhibits cooperative polyandry. In this case all males in the group copulate with the female and all participate in brood provisioning.

Polygynandry

Some mating systems have looser male-female bonds within groups. In polygynandrous groups, multiple females and males mate with each other, and males may care for the broods of several females. Chimpanzees and bonobos rely on this strategy it allows groups of males and females to live together and spend less time being concerned with mate competition. Polygynandry may be advantageous from the female's perspective because it causes paternity confusion, which decreases infanticide and allows her to have multiple males care for her brood.

Promiscuity

In promiscuity there are no pair bonds, and males and females, although sometimes choosy, often seem to mate randomly. As it is typically more advantageous for one or both sexes to pick their mate, promiscuity may occur in species for which the environment is unpredictable.

Sperm Competition

Although sperm competition is not a type of mating system per se, it is a form of male-male competition that plays an important role in mating systems. If more than one male mates with a female in a short time period, competition can occur after the males have released their sperm. In other words, once a male has released sperm, its sperm must be the first to reach an egg. This is often apparent in animals that use external fertilization. In aquatic animals that release their gametes into the water, animals that release the largest amount of sperm, and sperm that are highly capable of swimming, are likely to produce the most offspring. Animals with internal fertilization also experience sperm competition. Several mechanisms have evolved to facilitate a male's reproductive success with females that have multiple mates. For example, in one species of damselfly, males physically remove any sperm present from the female before it mates. Sperm competition adds to the difficulty of obtaining a successful reproductive event by males.

APPLICATION OF INBREEDING

Inbreeding is where the sire and dam have ancestors in common. The level of inbreeding is simply how closely related these relatives are across the sire and dam lines.

The standard (mathematical) measure for the level of inbreeding is the *Inbreeding Co-efficient.* It indicates the probability (between 0% and 100%) that genes at a randomly chosen location in the DNA are identical by descent. The technique assumes that there are 2 forms of a gene and that each form has an equal chance to be passed on to the next generation.

One limitation on calculating inbreeding coefficients is the depth of pedigrees available. Animals with a shallow pedigree may have a low inbreeding coefficient simply because their related ancestors are not on the database.

Understanding the Inbreeding Numbers

- Inbreeding Coefficient: is expressed as a percentage value. A low inbreeding coefficient means a low level of inbreeding. In most beef cattle breed societies, the vast majority of animals have an inbreeding coefficient of less than 10%, inbreeding coefficients over 30% are unusual, and over 40% are rare.
- Generations - Adjacent to the inbreeding coefficient, there are two numbers indicating the minimum number of generations in the animal's pedigree and, in brackets, the average number of generations in the animal's pedigree - as currently available on the database. Deeper pedigrees will result in a more accurate calculation of the inbreeding coefficient, particularly where the inbreeding coefficient is small.

Typical Inbreeding Coefficients

Typical inbreeding coefficients for various relationships are given in Table 5.5. The inbreeding coefficient will be higher if the ancestors in the pedigree are also inbred themselves

Table 5.5. Inbreeding coefficients for various inbred relationships

Relationship	**Inbreeding Coefficient (%)**
Animal mated to its own parent (eg Sire / daughter)	25%
Half sib matings (parents have a common sire or dam)	15.5%
Full sib matings (parents have a common sire and dam)	25%
Animal has a single common great grand parent	3.1%

Not surprisingly, smaller populations tend to have proportionally more animals with higher inbreeding coefficients than larger populations - simply because there are fewer candidate animals to select from (Figure).

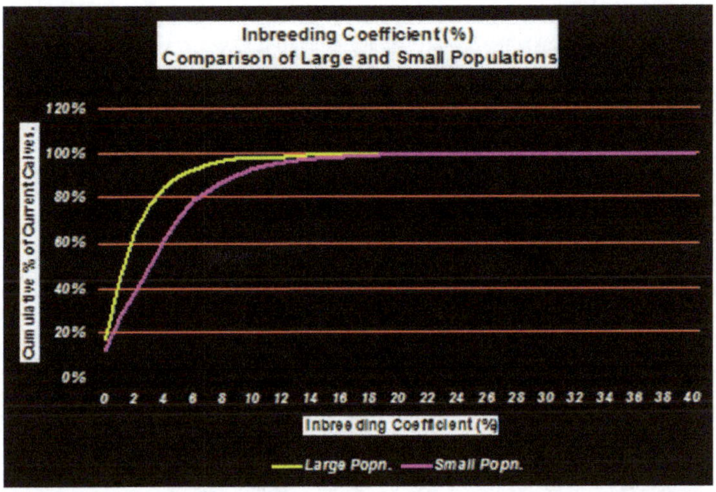

Figure . Comparison of Inbreeding Coefficients for large and small population size.

Discussion on Inbreeding

Selection in animal breeding systems uses genetic diversity/variation to improve the population by selecting superior animals for desirable traits. The more genetic variation observed in a population, the bigger the potential genetic gain possible in each generation. Conversely, decreased genetic variation increases the similarities of the population. Inbreeding reduces the amount of genetic diversity in a population.

Explanation of inbreeding in animal populations relies on a few basic genetic principles. Genetic information is stored in Chromosomes. Chromosomes are made up of DNA. Genes are sections of DNA and occur in pairs. A particular gene will occur at a particular site (locus, plural is loci) in the DNA of a particular Chromosome. The different forms of a gene (usually 2) that can occur at that locus are called alleles. Where both alleles are the same at the locus, they are called homozygous. Where the alleles are different, they are called heterozygous. In general, the two alleles will have an equal influence on the performance of an animal. That is, the heterozygous form (both alleles are present) tends to have performance midway between the two homozygous forms. In a few cases, one allele will have the main (dominant) effect on an animal, while the other allele will only have an effect in its homozygous form. These are called dominant and recessive genes.

The coefficient of inbreeding is the probability that two alleles at a randomly chosen locus are identical by descent. Note that alleles may be identical for other reasons, but the inbreeding coefficient is just looking at the mathematical probability that the alleles have come from a common ancestor.

Example of calculating an Inbreeding Coefficient for an animal that has a common great grandparent:

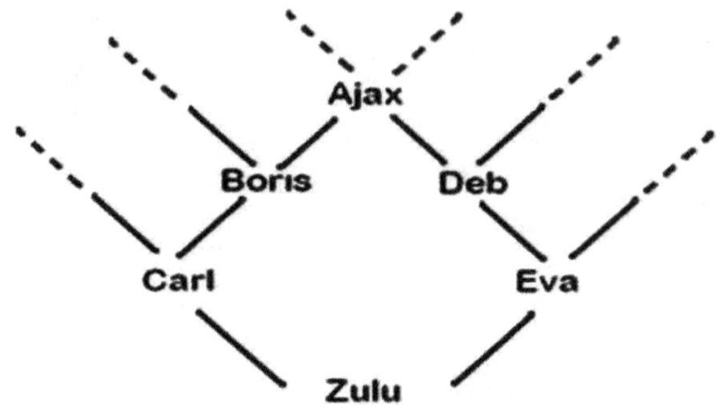

Zulu is inbred because Ajax is a common ancestor.
There are 5 ancestors in the inbreeding loop:
Carl - Boris - Ajax - Deb - Eva

Zulu's Inbreeding Coefficient = $(1/2)^5 = 1/32 = 3.1\%$

(If any of Zulu's ancestors were themselves inbred, then Zulu's inbreeding coefficient would be higher)

As inbreeding increases, the frequency of alleles being homozygous at a particular locus also increases. Hence inbreeding reduces the amount of variation in a population.

Line breeding attempts to maximize desirable homozygous alleles (ie fix these desirable genes in the population) using selection and planned breeding programs – a sort of controlled inbreeding. However, an increase in homozygous alleles may also increase homozygous alleles with undesirable effects which can result in reduced performance and/or fertility (inbreeding depression); or even be fatal.

The converse of inbreeding is out-crossing. Some component of observed hybrid vigor is the out-crossing effect. Note that two inbred animals that are unrelated themselves will have fully out crossed progeny (inbreeding coefficient of zero). Hence, you can go from inbred to out crossed in one generation.

Inbreeding Depression

Inbreeding depression is a term used to describe the reduction in performance and viability due to the increase in inbreeding levels (reduced genetic variation). Reproductive fitness tends to be affected more than performance traits in this respect. The extent of inbreeding depression depends on the population being measured, the level of genes already fixed in the population, the frequency of any deleterious genes that may exist in the population and chance effects of which particular genes are affected.

There is no defined limit as to what is an acceptable level of inbreeding in domestic animal populations. However, inbreeding depression is likely to be more apparent once inbreeding levels get to above 10%. As a very rough guide, there is often a 2-20% decrease in performance of the trait per 10% of inbreeding coefficient.

Undesirable and Lethal Recessives

Some undesirable recessive genes are only expressed in their homozygous form. In these cases, the dominant, desirable allele of the gene is expressed in the animal for both the dominant homozygous and heterozygous forms.

The Arthrogryposis Multiplex (AM) gene is an example of a recessive gene that is lethal. The undesirable allele (a) in its homozygous form (aa) causes the affected animal to have a bent and twisted spine and limited muscle development, but with apparent normal brain tissue. Affected calves are generally still born. Conversely, animals with at least one copy of the dominant allele (A) are normal, unaffected calves (ie the homozygous AA or heterozygous Aa forms of the gene). The heterozygous animals (Aa) are called carriers as they have the potential to pass on the undesirable form (a) of the gene undetected to the next generation.

A carrier (Aa) animal mated to a homozygous normal (AA) animal will result in a 50% chance of the progeny being homozygous normal (AA) and 50% chance of being a heterozygous carrier (Aa) with no lethal recessive (aa) forms (Table 5.3).. However, a carrier (Aa) mated to another carrier (Aa) has a 25% chance of producing homozygous normal (AA) offspring, 50% chance of producing carrier (Aa) offspring and 25% chance of producing affected (aa) offspring - which die at birth

Table 5.3: Probability of mating outcomes of heterozygous carriers and homozygous non-carriers

Mating		**Non-carrier (AA)**		**Carrier (Aa)**	
outcomes		A	A	A	a
Carrier	A	AA	AA	AA	Aa
(Aa)	a	Aa	Aa	Aa	aa
Outcomes:		AA 50%, Aa 50%		AA 25%, Aa 50%, aa 25%	

A genetic test is now available to detect which alleles the animal actually has - ie whether an animal is homozygous for the desirable allele (AA) or is a carrier (Aa).

APPLICATION OF CROSSBREEDING

Improvement of the economic position of the farm or ranch is an ongoing process for many commercial cow-calf producers. Profitability may be enhanced by increasing the volume of production (i.e. the pounds of calves you market) and/or the value of products you sell (improving quality). The reduction of production costs, and thus breakeven prices, can also improve profitability. More and more producers are finding that a structured crossbreeding system helps them achieve the goals of increasing productivity and reducing production costs. Indeed, pricing differences, popularity and perceptions of utility of some breeds and color pattern have motivated producers to stray away from sound crossbreeding systems.

Why Crossbreed?

The use of crossbreeding offers two distinct and important advantages over the use of a single breed. First, crossbred animals have heterosis or hybrid vigor. Second, crossbred animals combine the strengths of the parent breeds. The term 'breed complementarity' is often used to describe breed combinations that produce highly desirable animals for a broad range of traits.

Heterosis

Heterosis, also called hybrid vigor, occurs when the performance of crossbred progeny is different (usually better) than the average of their parent types, as shown in the example in Figure . This example illustrates direct heterosis, the effect of a crossbred individual's gene combinations on its performance. There can also be maternal heterosis, or the indirect effect of a crossbred dam's gene combinations that influence her calf's performance through the maternal environment she provides.

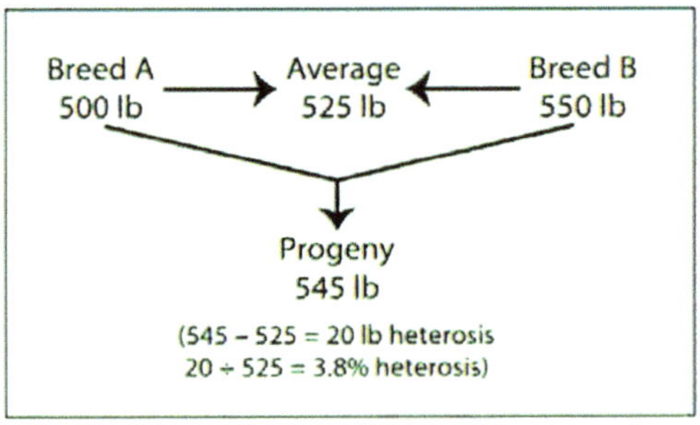

Figure . Heterosis.

Heterosis is the opposite of inbreeding depression, so it is greatest in the progeny of least related parents. For instance, there is greater heterosis in crossing the less related breeds Hereford and Brahman than in crossing the more closely related breeds Hereford and Angus. Heterosis is reduced when both parents contain the same breed. There is no loss of heterosis if the parents share no breed in common, whether the parents are purebreds or crossbreds. If an F_1 is bred to either of its parent breeds (called a backcross), heterosis in its progeny is about half of that expressed in the F_1. If the F_1 is not backcrossed but is bred to a third breed, or an unrelated crossbred, there should be no loss of heterosis in its progeny because the sire and dam have no breed in common as in a backcross.

If two F_1s of the same cross are mated, the progeny, called F_2, should have about half the heterosis of the F_1. However, if F_2s and subsequent generations of this cross are intermated, their progeny average no additional loss of heterosis except for any inbreeding that might develop over time. Characteristics express heterosis differently. Heterosis tends to be highest in fertility, livability, and longevity; intermediate in milk production, weight gain, feed efficiency, and body size; and lowest in carcass traits. Even with the benefit of heterosis the crosses from mediocre parents usually do not perform as well as superior purebreds. So, the most productive crossbreds come from genetically superior parents.

Types of Heterosis

Heterosis is of two types:

1. True heterosis

It is inherited. It can be further divided into two types:

- **Mutational true heterosis:** It is the sheltering or shadowing of the deleterious, unfavorable, often lethal, recessive mutant genes by their adaptively superior dominant alleles.
- **Balanced true heterosis:** It arises out of balanced gene combinations with better adaptive value and agricultural usefulness.

5. Pseudo-heterosis

Crossing of the two parental forms brings in an accidental, excessive and un-adaptable expression of temporary vigor and vegetative overgrowth. It is also called luxuriance.

Causes of Heterosis

The phenomenon of heterosis can be explained on the basis of the causes: Genetic causes and Physiological causes.

Genetic Causes

There are two possible causes of heterosis viz.;

- Dominance
- Over-dominance

Dominance hypothesis:

This theory was proposed by Davenport (1910), Bruce (1910) and Keable and Pellew (1910).This theory is based on the assumption that hybrid vigor results from bringing together female dominant genes. According to this theory, genes that are favorable for vigor and growth are dominant, and genes that are harmful to the individual are recessive.

The dominant genes contributed by one parent may complement the dominant genes contributed by the other parent, so that F_1 will have the more favorable combination of dominant genes, than either parent e.g., Dominant genes ABCDE are favorable for good yield. Inbred A has the genotype AA BB cc dd (AB dominant) and inbred B has the genotye aa bb CC DD (CD dominant).

The genotype of the F_1 **hybrid is as follows:**

Inbred A	×	Inbred B
AA BB cc dd		aa bb CC DD

F1 hybrid
Aa Bb Cc Dd

The F_1 hybrid contains dominant genes at all the loci represented here (ABCD) and exhibits more vigor than either of the parent inbred lines.

Objections:

There are two main objections:

- **Failure in the isolation of inbreeds as vigorous as hybrids:** If the above hypothesis is true, it should be possible to isolate inbreeds with all dominant genes. Such inbreeds would be as vigorous as the F_1 hybrids. However, such hybrids have not been isolated. Jones in this modified theory entitled "Dominance of Linked Genes Hypothesis" provided explanation for this.
- He suggested that there may be a linkage between some favorable dominant genes and some un-favorable recessive genes and as a result it is not possible to obtain true breeding homozygous individual for all dominant genes in F_2 generation.
- **Symmetrical distribution in F_2:** It is already studied that in F_2 dominant and recessive characters segregate in the ratio of 3:1. If heterosis is due to dominance of independent factors, the F_2 distribution curve for heterotic character should be skewed (slanting) rather than smooth and symmetrical.
- But the curve of F_2 is always, smooth and symmetrical and not skewed. This objection was removed by Collins. According to him trait like yield is governed by large number of genes or polygenes, which exhibit continuous variation resulting in symmetrical distribution of genes.

Over dominance hypothesis

This hypothesis was given by Shull and East independently. According to the supposition hybrid vigor on the basis of heteozygosity is superior to homozygosity. According to this hypothesis there are contrasting alleles for example a_1 and a_2, for a single locus. Each allele produces favorable yet different effects in the plant.

In a heterozygous plant (a_1, a_2) a combination of the effects is produced which is more favorable in the plant than the effect produced by either of the alleles alone. This phenomenon of heterozygote (a_1 a_2) being superior to the homozygotes (a_1a_1 or a_2a_2) is termed over dominance.

Various names have been given to this idea e.g., super dominance, interaction of alleles at a single locus over-dominance etc., but the term over-dominance is widely accepted.

Breed Differences and Complementarity

The amount of variability between breeds for most traits is comparable to the amount of variability one would expect to find between individuals within a breed. All breeds manifest superiority in some of the economically important traits, but no breed can boast excellence in all traits. Complementarity helps match the genetic potential for economically important traits such as growth rate and carcass composition with climate, feed resources and market preferences. Simply put, breed complementarity means that the strengths of one breed can complement or mask the weaknesses of another breed. In practical terms, a producer looking to utilize complementarity would choose a bull breed that would pass on rapid growth and desirable carcass traits to breed to his crossbred cows that would provide adequate milk for the rapidly growing calf and produce a live, healthy calf each year.

In poorly conceived crossbreeding programs, complementarity could have negative effects on productivity. For example, if a large, terminal sire breed was bred to small, immature or "hard-calving" cows, the result could be an increase in dystocia problems. Cattle breeds can be separated into different biological types, with each type exhibiting differing levels of production for various production characteristics. Table 5.4 lists some breeds grouped by biological type.

Table 5.4: Cattle breeds grouped by biological type.

Breed	Milk Production	Growth Rate and Mature Size	Percentage Retail Product	Age at Puberty
Jersey	*****	*	*	*
Herford	***	**	*	***
Angus	***	**	*	**
Brahman	***	***	***	*****
Tarentaise	****	***	****	**
Simmental	****	*****	*****	**
Gelbvich	****	****	****	*
Main Anjou	**	*****	****	**
Limousin	*	***	*****	****
Charolais	**	*****	*****	****
Chiania	**	*****	*****	****

One extreme crossbreeding example which demonstrates breed differences and complementarity is a scheme which was popular in some areas of the country in the 1970's. A Jersey bull would be crossed onto Angus cows to produce medium frame, high milking F_1 females. These were then crossed with Charolais bulls to produce terminal calves. The Jersey provided the genes for milk production and marbling ability; the Angus, the genes for carcass quality; and the Charolais, the genes for superior growth.

Crossbreeding Systems

Crossbreeding systems implemented in a commercial herd vary considerably from herd to herd. A number of factors determine the practicality and effectiveness of crossbreeding systems for each operation. These factors include herd size, market target, existing breeds in the herd, the level of management expertise, labor availability, grazing system, handling facilities and the number of available breeding pastures. It should be noted that in some instances the number of breeding pastures required can be reduced through the use of artificial insemination. Additional considerations include the operator's decision to purchase replacement females or select and raise replacements from the herd. Purchasing healthy, well developed replacement females of appropriate breed composition can be the simplest and quickest way for producers, especially small operators, to maximize maternal heterosis in the cowherd. Regardless of the crossbreeding system selected, a long-term plan and commitment to it are required to achieve the maximum benefit from crossbreeding. A variety of crossbreeding systems are described on the following pages. These systems are summarized in Table 5.5 by their productivity advantage measured in percentage of pounds of calf weaned per cow exposed. Additionally the table includes the expected amount of retained heterosis, the minimum number of breeding pastures required, whether purchased replacements are required, the minimum herd size required for the system to be effectively implemented, and the number of breeds involved.

Table 5.5: Summary of crossbreeding systems by amount of advantage and other factors.

Type of System		% of Cow Herd	% of Marketed Calves	Advantage (%)[b]	Retained Heterosis (%)[c]	Minimum No. of Breeding Pastures	Minimum Herd Size	No. of Breeds
2-Breed Rotation	A*B Rotation	100	100	16	67	2	50	2
3-Breed Rotation	A*B*C Rotation	100	100	20	86	3	75	3
2-Breed Rotational/ Terminal Sire	A*B Rotational	50	33			2		
	T x (A*B)	50	67			1		
	Overall	100	100	21	90	3	100	3
Terminal Cross with Straightbred Females[d]	T x (A)	100	100	8.5	0[e]	1	Any	2
Terminal Cross with Purchased F_1 Females	T x (A*B)	100	100	24	100	1	Any	3
Rotate Bull every 4 years	A*B Rotation	100	100	12-16	50-67[f]	1	Any	2
	A*B*C Rotation	100	100	16-20	67-83[f]	1	Any	3
Composite Breeds	2-breed	100	100	12	50	1	Any	2
	3-breed	100	100	15	67	1	Any	3
	4-breed	100	100	17	75	1	Any	4
Rotating Unrelated F_1 Bulls	A*B x A*B	100	100	12	50	1	Any	2
	A*B x A*C	100	100	16	67	1	Any	3
	A*B x C*D	100	100	19	83	2	Any	4

Two-Breed Rotation

A two-breed rotation is a simple crossbreeding system requiring two breeds and two breeding pastures. The two-breed rotational crossbreeding system is initiated by breeding cows of breed A to bulls of breed B. The resulting heifer progeny (A*B) chosen as replacement females would then be mated to bulls of breed A for the duration of their lifetime. Note the service sire is the opposite breed of the female's own sire. These progeny are then ¼ breed A and ¾ breed B. Since these animals were sired by breed B bulls, breeding females are mated to breed A bulls. Each succeeding generation of replacement females is mated to the opposite breed of their sire. The two-breed rotational crossbreeding system is depicted in Figure . Initially only one breed of sire is required. Following the second year of mating, two breeds of sire are required. After several generations the amount of retained heterosis stabilizes at about 67% of the maximum calf and dam heterosis, resulting in an expected 16% increase in the pounds of calf weaning weight per cow exposed above the average of the parent breeds. This system is sometimes called a crisscross.

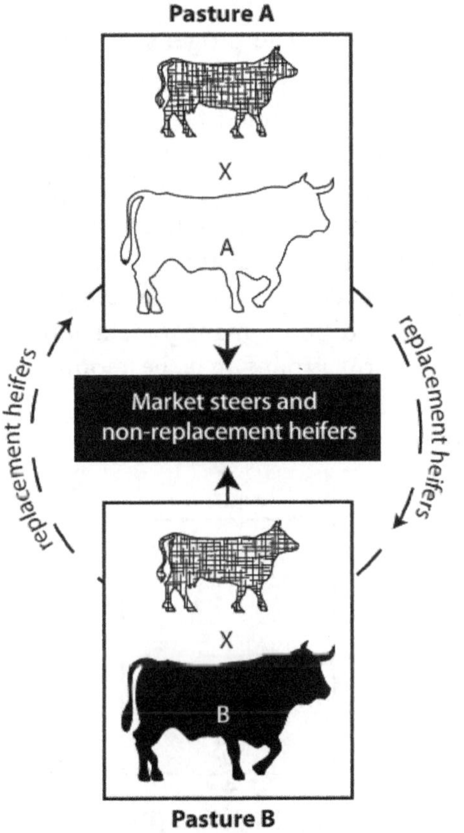

Figure . Two-breed rotation.

Requirements

A minimum of two breeding pastures are required for a two-breed rotational system if natural service is utilized exclusively. Replacement females must be identified by breed of sire to ensure proper matings. A simple ear tagging system may be implemented to aid in identification. All calves sired by breed A bulls should be tagged with one color (e.g. red) and the calves sired by bulls of breed B should be tagged with a different color (e.g. blue). Then at mating time, all the cows with red tags (sire breed A) should be mated to breed B bulls, and vice-versa.

Considerations

The minimum herd size is approximately 50 cows with each half being serviced by one bull of each breed. Scaling of herd size should be done in approximately 50 cow units to make the best use of service sires, assuming 1 bull per 25 cows. Replacement females are mated to herd bulls in this system so extra caution is merited in sire selection for calving ease to minimize calving difficulty. Be sure to purchase bulls or semen from sires with acceptable Calving Ease (preferably) or Birth Weight EPDs for mating to heifers. Alternately, a calving ease sire(s) could be purchased to breed exclusively to first calf heifers regardless of their breed type. The progeny produced from these matings that do not conform to the breed type of the herd should all be marketed.

Breeds used in rotational systems should be of similar biological type to avoid large swings in progeny phenotype due to changes in breed composition. The breeds included have similar genetic potential for calving ease, mature weight and frame size, and lactation potential to prevent excessive variation in nutrient and management requirements of the herd. Using breeds of similar biological type and color pattern will produce a more uniform calf crop which is more desirable at marketing time. If animals of divergent type or color pattern are used, additional management inputs and sorting of progeny at marketing time to produce uniform groups may be required.

Three-Breed Rotation

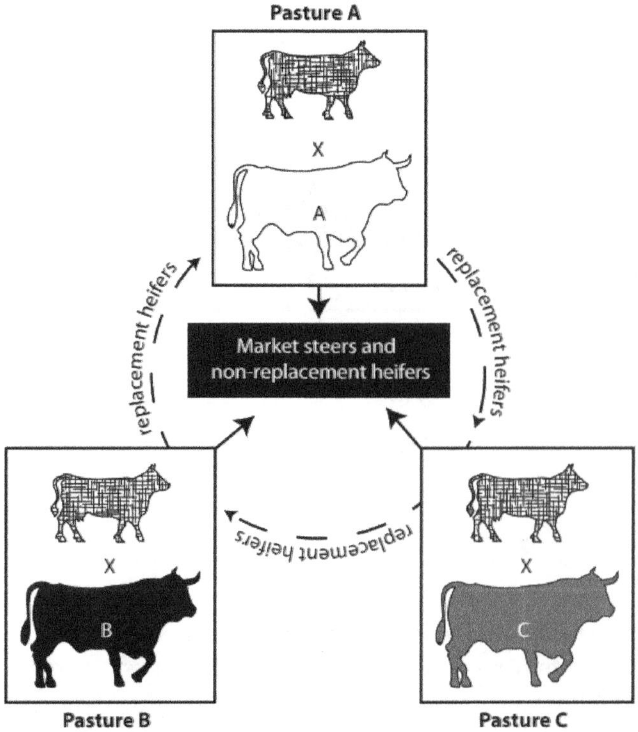

Figure : Three-breed rotation.

A three-breed rotational system is very similar to a two-breed system in implementation with an additional breed added to the mix. This system is depicted in Figure . A three-breed rotational system achieves a higher level of retained heterosis than a two-breed rotational crossbreeding system does. After several generations the amount of retained heterosis stabilizes at about 86% of the maximum calf and dam heterosis, resulting in an expected 20% increase in the pounds of calf weaning weight per cow exposed above the average of the parent breeds. Like the two-breed system, distinct groups of cows are formed and mated to bulls of the breed which represents the smallest fraction of the cows breed makeup. A cow will only be mated to a single breed of bull for her lifetime.

Requirements

A minimum of three breeding pastures are required for a three-breed rotational system. Replacement females must be identified by breed of sire to ensure proper matings. A simple ear tagging system may be implemented to aid in identi- fication. All calves sired by breed A bulls should be tagged with one color (e.g. red), the calves sired bulls of breed B should be tagged with a different color (e.g. blue), and the progeny of bulls of breed C tagged a third color (e.g. green). Then at mating time, all the cows with red tags (sired by breed A) should be mated to breed B bulls, cows with blue tags (sired by breed B) should be mated to breed C bulls, and, finally, all cows with green tags (sired by breed C) should be mated to breed A bulls.

Considerations

The minimum herd size is approximately 75 cows with each half being serviced by one bull of each breed. Scaling of herd size should be done in approximately 75 cow units to make the best use of service sires, assuming 1 bull per 25 cows. Replacement females are mated to herd bulls in this system so extra caution is merited in sire selection for calving ease to minimize calving difficulty. Be sure to purchase bulls or semen from sires with acceptable Calving Ease (preferably) or Birth Weight EPDs for mating to heifers. Alternately, a calving ease sire(s) could be purchased to breed exclusively to first calf heifers regardless of their breed type. The progeny produced from these matings that do not conform to the breed type of the herd should all be marketed.

Breeds used in rotational systems should be of similar biological type to avoid large swings in progeny phenotype due to changes in breed composition. The breeds included have similar genetic potential for calving ease, mature weight and frame size, and lactation potential to prevent excessive variation in nutrient and management requirements of the herd. Using breeds of similar biological type and color pattern will produce a more uniform calf crop which is more desirable at marketing time. If animals of divergent type or color pattern are used additional management inputs and sorting of progeny at marketing time to produce uniform groups may be required.

2-Breed Rotational/Terminal Sire

The two-breed rotational with terminal sire system is sometimes called a rota-terminal system. It includes a two-breed rotational crossbreeding system of maternal breeds A and

B. This portion of the herd is charged with producing replacement females for the entire herd, so maternal traits of the breeds included are very important. The remainder of the cow herd is bred to a terminal sire of a different breed as illustrated in Figure . In this system approximately half of the cow-herd is committed to the rotational portion of the breeding system and half to the terminal sire portion. This system retains about 90% of the maximum calf heterosis plus capitalizes on 67% of the maximum dam heterosis; it should increase weaning weight per cow exposed by approximately 21%.

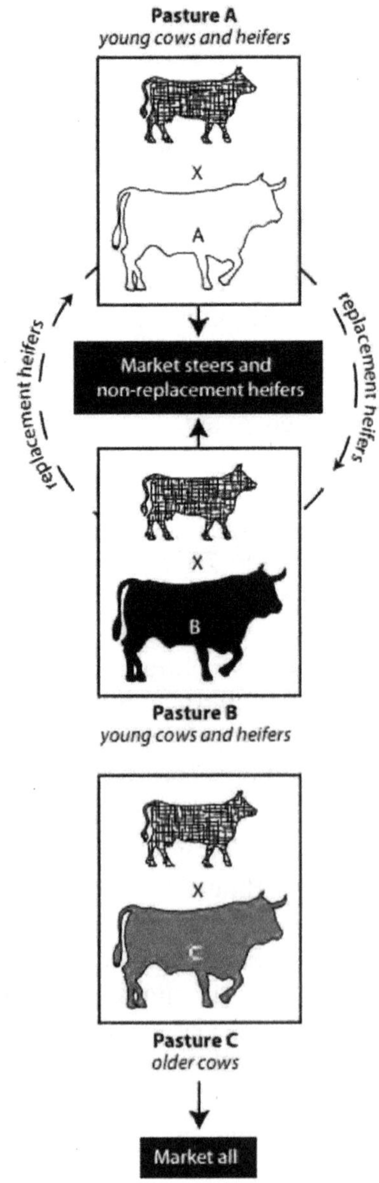

Figure . Two-breed rotational/terminal sire.

Requirements

This system requires a minimum of three breeding pastures. Females in the rotational portion of the system must be identified by breed of sire. Minimum herd size is approximately 100 cows. Given the complexity of the breeding system and identification requirements, this system requires more management and labor to make it run effectively than some other systems do. The trade off in systems that are easier to manage is that they typically yield lower levels of heterosis. If management expertise and labor are readily available this system is one of the best for maximizing efficiency and the use of heterosis.

Considerations

The females in the rotational portion should consist of the youngest females, namely the 1-, 2-, and 3-year-olds. These females should be bred to bulls with both good calving ease and maternal traits. Calving ease and maternal traits are emphasized here because the cows being bred are the youngest animals where dystocia is expected to be highest. Additionally, replacement females for the entire herd will be selected from the progeny of these cows so maternal traits are important. The remainder of the cow herd consists of mature cows that should be mated to bulls from a third breed that excel in growth rate and muscularity.. When fewer replacements are needed a smaller portion of the herd will be included in the rotational system. Be sure to keep the very youngest breeding females in the rotational system to avoid dystocia problems. If ownership of calves will be retained through harvest some consideration should be given to end product traits such as carcass weight, marbling, and leanness. One drawback of the system is that there will be two different types of calves to market: one set from the maternally focused rotational system and one from the terminal sire system. Sorting and marketing can typically help offset this problem. The benefits of the rota-terminal system are usually worth the limitations.

Two-Breed Terminal Sire

A two breed terminal cross system uses straightbred cows of one breed and a sire(s) of another breed. No replacement females are kept and therefore, must be purchased. Since all calves are marketed it is a terminal sire system. Charolais or Limousin sires used on Angus cows would be a common example. Implementations of two breed terminal sire systems are not desirable or recommended as they do not employ any benefits of maternal heterosis as the cows are all straightbred. Remember most of the benefits of heterosis arise from the enhancement of reproduction and longevity traits of crossbred cows. A slight improvement in pounds of calf weaned per cow exposed will be observed due to individual heterosis in the calves produced by this system.

Terminal Cross with Purchased F_1 Females

The terminal cross system utilizes crossbred cows and bulls of a third breed as shown in Figure . This system is an excellent choice as it produces maximum heterosis in both the calf and cow. As such, calves obtain the additional growth benefits of hybrid vigor while heterosis in the cows improves their maternal ability. The terminalcross system is one of the simplest systems to implement and achieves the highest use of heterosis and breed complementarity. All calves marketed will have the same breed composition. A 24% increase in pounds of calf weaned per cow exposed is expected from this system when compared to the average of the parent breeds.

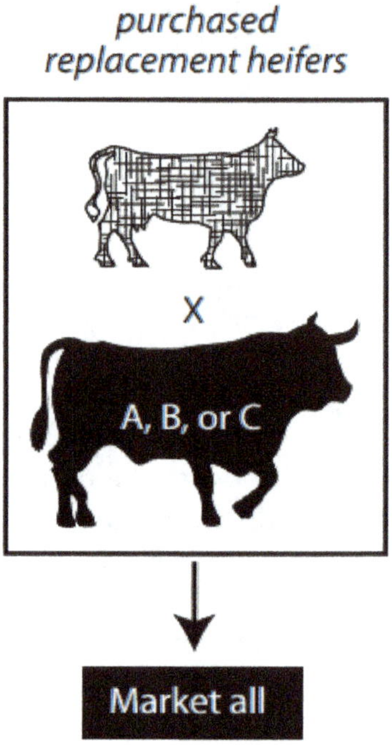

Figure . Terminal cross with purchased F_1 females.

Requirements

The terminal cross system works well for herds of any size if high quality replacement females are readily available from other sources. Only one breeding pasture is required. No special identification of cows or groups is required.

Considerations

Since replacement females are purchased care should be given in their selection to ensure that they are a fit to the production environment. Their adaptation to the production environment will be determined by their biological type, especially their mature size and lactation potential. Success of the system is dependent on being able to purchase a bull of a third breed that excels in growth and carcass traits. If virgin heifers are selected as replacements, they should be mated to an easy calving sire to minimize dystocia problems. Alternately, three year-old cows may be purchased as replacements and mated to the terminal sire breed. Disease issues are always a concern when introducing new animals to your herd. Be sure that replacement heifers are from a reputable, disease-free source and that appropriate biosecurity measures are employed. Johnes, brucellosis, tuberculosis, bovine viral diarrhea (BVD) are diseases you should be aware of when purchasing animals. Another consideration and potential advantage of the terminal-cross system is that replacement females do not need to be purchased each year depending on the age stratification of the original cows. In some cases replacements may be added every 2-5 years providing an opportunity to purchase heifers during periods of lower prices or more abundant supplies. Heifers could also be developed by a professional heifer development center or purchased bred to easy calving bulls.

Rotate Bull Every Four Years

This system requires the use of a single breed of sire for four years then a rotation to a second breed for four years, then back to the original breed of sire for four years, and so on. This system is depicted in Figure . Breed fractions of cows and level of maternal heterosis will vary depending on sequence of production. Estimates of the range of retained heterosis are dependant on the number and breed make-up of females retained in the herd. Several assumptions are made when estimating the expected performance improvement and retained heterosis. In a two-breed rotation of bulls the minimum retained heterosis is 50% and assumes that over time the average breed fractions represented in the herd are equal (50% breed A, 50% breed B) with random selection of replacement females. However, depending on culling rate and replacement selection, this retained heterosis maybe as high as 67%, similar to a true two-breed rotation. The expected improvement in weaning weight per cow exposed is a function of retained heterosis will range from 12-16% for at two breed system with bulls rotated every four years.

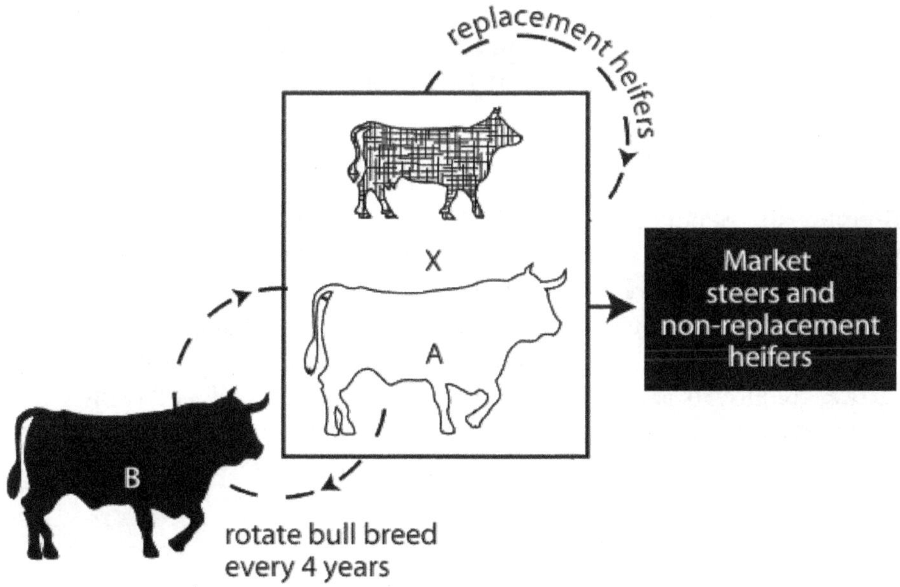

Figure . Rotate bull every four years.

Likewise, in a three-breed rotation of bulls every four years, the minimum expectation of retained heterosis is 67% assuming the animals stabilize at a composition of 1/3 of each breed. Again, depending on culling rate and replacement selection the retained heterosis may be as high as 83% which is similar to a true 3 breed rotational system. The expected improvement in weaning weight per cow exposed is a function of retained heterosis will range from 16-20% for at three breed system with bulls rotated every four years.

Requirements

The rotate bulls every four years system is particularly useful for small herds or herds with minimal management or labor inputs as only one breeding pasture is required and cows are not required to be identified by breed of sire. Replacement females are kept in this system but should only be kept from the first two calf crops of a bull breed cycle. Some sire-daughter matings will occur in this system during years three and four of a sire breed cycle. Siredaughter matings increase inbreeding and over represents the breed of sire in the resulting calves. Both decrease heterosis and these calves desirability as replacement females. Bulls may be replaced after two breeding seasons to minimize sire-daughter matings. This strategy, however, make less efficient use of capital investments in bulls given their useful life is longer than two years. This decreased efficiency has to be balanced against the limitation of retaining replacements during two of every four years in a sire-breed cycle. This limitation may be of little consequence in small herds, but large fluctuations in cow inventory may result if this system is utilized in large operations

Considerations

This system does not maximize heterosis retention, but it is very simple to implement and manage. The first breed of sire should be used for five calf crops if you start with straightbred cows to optimize retention of heterosis.

Composite Breeds

The use of composite populations in beef cattle has seen a surge in popularity recently. Aside from the advantages of heterosis retention and breed complementarity, composite population breeding systems are as easy to manage as straightbreds once the composite is formed. The simplicity of use has made composites popular among very large, extensively managed operations and small herds alike. When two-, three- or four-breed composite are formed they retain 50%, 67%, and 75% of maximum calf and dam heterosis and improve productivity of the cowherd by 12%, 15%, and 17%, respectively. Thus, these systems typically offer a balance of convenience, breed complementarity and heterosis retention. A composite breeding system is presented in Figure 0.

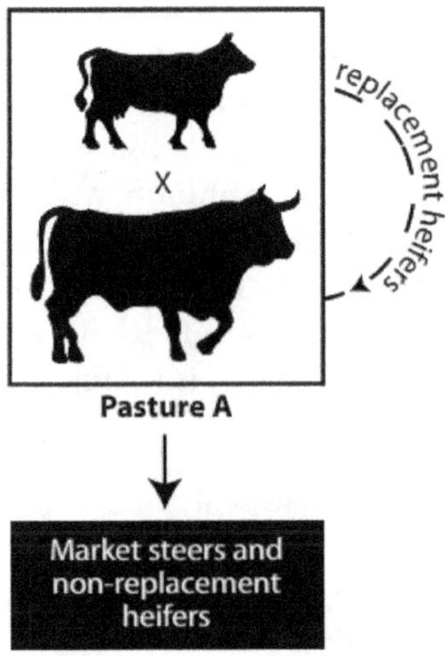

Figure 0: Composite breeding system.

Requirements

This requires either a very large herd (500 to 1000 cows) to form your own composite or a source of composite bulls or semen. In closed populations inbreeding must be avoided as it will decrease heterosis. To help minimize inbreeding in the closed herd where cows are randomly mated to sires. Similar recommendations would be made to seedstock breeders wishing to develop and merchandize bulls of a composite breed. In small herds, inbreeding may be avoided through purchase of outside bulls that are unrelated to your herd. Due to the ease of use once the composite is established, composite systems can be applied to herds of any size or number of breeding pastures.

Considerations

Clearly, availability of outside seedstock is the limiting factor for most producers. However, with emerging popularity of structured, stabilized half-blood systems (inter se mated F1 animals) such as SimAngus, Balancer and LimFlex, availability is much easier for these British x Continental crossbreds. Other composites have been formed and include: MARC I, MARC II, MARC III, Rangemaker, Stabilizer, and others.

Rotating Unrelated F_1 Bulls

The use of F_1, or first cross, bulls resulting from the cross of animals from two breeds is becoming more wide spread. F_1 bulls provide a simple alternative to the formulation of composite breeds. Additionally, the F_1 systems may provide more opportunity to incorporate superior genetics as germplasm can be sampled from within each of the large populations of purebreds rather than a smaller composite population. The use of unrelated F1 bulls, each containing the same two breeds, in a mating system with cows of the same breeds and fractions will result in a retention of 50% of maximum calf and dam heterosis and an improvement in weaning weight per cow exposed of 12%. A system that uses F_1 bulls that have a breed in common with the cow herd (A*B x A*C) results in heterosis retention of 67% and an expected increase in productivity of 16%. While the use of F_1 bulls that don't have breeds in common with cows made up of equal portion of two different breeds (A*B x C*D) retains 83% of maximum heterosis and achieves productivity gains of 19%. This last system is nearly equivalent to a three breed rotational system in terms of heterosis retention and productivity improvement, but much easier to implement and manage. These three systems are depicted in Figure 1.

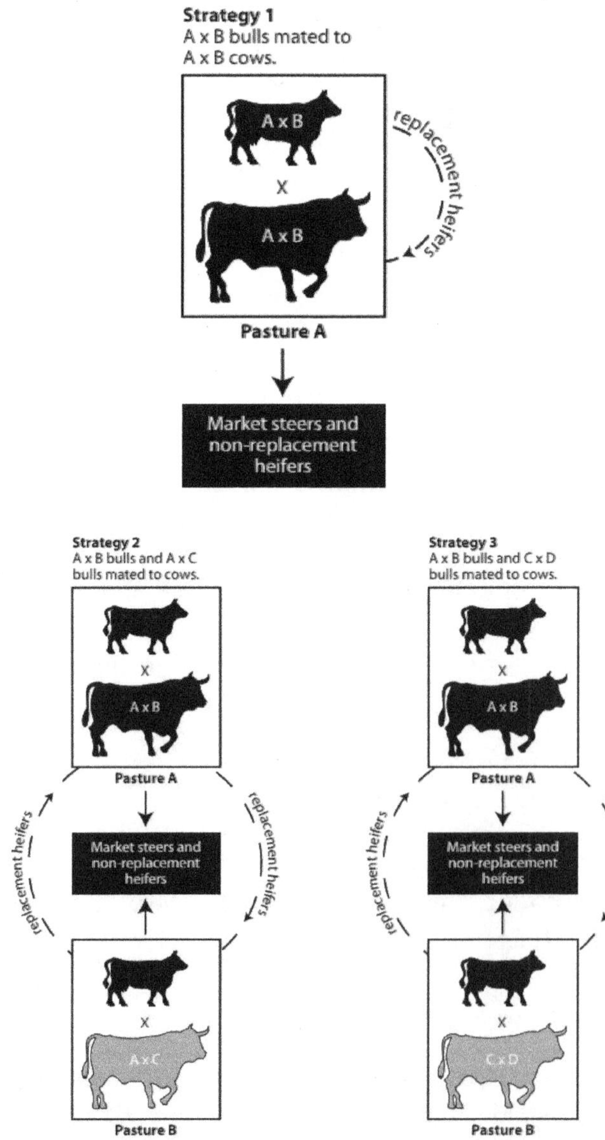

Figure 1: Rotating F_1 bulls.

Requirements

The use of F_1 bulls requires a seedstock source from which to purchase. The bulls will need to be of specific breed combinations to fit your program. These programs fit a wide range of herd sizes. The use of F_1 bulls on cows of similar genetic makeup is particularly useful for small herds as they can leverage the power of heterosis and breed complementarity using a system that is as simple as straight breeding. Additionally, operators of this system can keep their own replacement females.

Considerations

The inclusion of a third or fourth breed in the systems takes more expertise and management. To prevent wide swings in progeny phenotype, breeds B and C should be similar in biological type, while breeds A and D should be similar in biological type.

Crossbreeding Challenges

Although crossbreeding has many advantages, there are some challenges to be aware of during your planning and implementation as outlined by Ritchie et al., 1999.

- More difficult in small herds—Crossbreeding can be more difficult in small herds. Herd size over 50 cows provides the opportunity to implement a wider variety of systems. Small herds can still benefit through utilization of terminal sire, composite or F_1 systems.
- Requires more breeding pastures and breeds of bulls—Purchasing replacements and maximum use of A.I. can reduce the number of pastures and bulls. However, most operations using a crossbreeding system will expand the number of breeding pastures and breeds of bulls.
- Requires more record keeping and identification of cows—Cow breed composition is a determining factor in sire breed selection in many systems.
- Matching biological types of cows and sire—Breed complementarity and the use of breed differences are important advantages of cross breeding. However, to best utilize them care must be given in the selection of breeds and individuals that match cows to their production environment and sires to market place. Divergent selection of biological type can result in wide swings in progeny phenotype in some rotational systems. These swings may require additional management input, feed resources, and labor to manage as cows or at marketing points.
- System continuity—Replacement female selection and development is a challenge for many herds using crossbreeding systems. Selection of sires and breeds for appropriate traits (maternal or paternal traits) is dependent of ultimate use of progeny. Keeping focus on the system and providing labor and management at appropriate times can be challenging. Discipline and commitment are required to keep the system running smoothly.

BREEDS OF CATTLE AND BUFFALO

A breed is a group of animals related similar characters like general appearance. Size, features and configuration etc. A group of individual which have certain common characteristics that distinguish them from other groups of individuals is known as "Species".

Cattle Breeds

Holstein

The most popular of all the dairy cow breeds, Holsteins are black and white (and sometimes red). Their markings are like human fingerprints: no Holsteins have the same markings. Holsteins are the most common dairy cow because they tend to produce more milk than all the other breeds.

Jersey Cow

Jersey cows, named for the British isle of Jersey where they originated, are most often brown or tawny with a light-colored underbelly and dark hooves. Their milk is the richest of all the dairy cows and high in butterfat, making it the ideal milk for the production of butter and cheese.

Guernsey

Also known as the Royal Breed because of the golden hue of their milk. The golden tone is due to an exceptionally high content of beta carotene, a source of Vitamin A. Guernseys themselves are also a golden color and can have white markings. As their name suggests, Guernseys were originally bred on the British Channel Island of Guernsey.

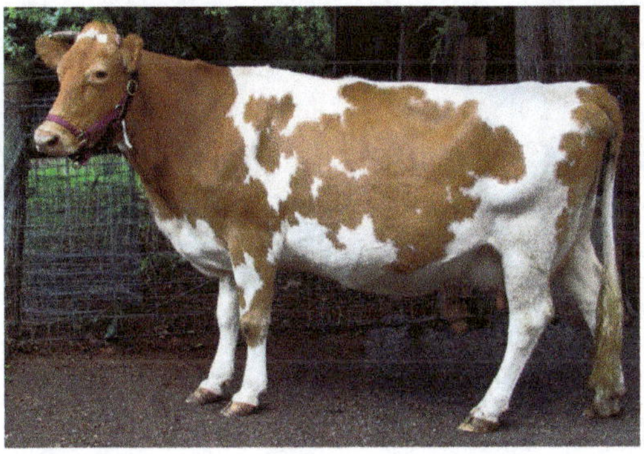

Ayrshire

Ayreshire cows originated from Ayr in the southwest of Scotland. They were first imported into Australia in the 1850's. The breed's coat color can vary greatly from any shade of red or brown with white or mahogany. They are characterized as having a medium to high milk composition and medium to high milk yield. There are only few whole Ayreshire herds being commercially milked in Australia.

Brown Swiss

This breed native to Switzerland (if you needed us to spell it out) is one of the oldest in existence. Their color varies from light-to-dark brown and sometimes gray, but they are easily recognized by their large furry ears. Brown Swiss have a very gentle nature.

Milking Shorthorns

Once known as Durhams, these cows were imported from Northeastern England. Washington's dairy farming tradition actually began in 1836 when Dr. Marcus Whitman brought 16 Durham cows to serve the mission he established in Walla Walla. Milking shorthorns can be red, red with white marking, white or roan.

Dutch Belted

A Dutch belted cow may be the easiest dairy cow to identify because of the wide "belt" of white around its middle section. It is estimated that there are less than 300 Dutch Belted cattle in the United States.

Each breed is a different size, shape and color; but now that you've met our dairy cow family, you should be able to tell them apart when you see them grazing in a nearby field.

Buffalo Breeds

Water Buffalo

The water buffalo (Bubalus bubalis) is the largest buffalo and in fact the largest of all members of family Bovidae. Weighing between 1,500 and 2,650 pounds, and standing between 5 and 6.2 feet at the shoulder, the water buffalo is formidable. Males of the species have horns, which can span up to 5 feet across. They live in wetlands and marshes in the tropical or subtropical forests of Asia. The species are currently endangered and live in small, protected areas of Nepal, India, Bhutan and Thailand.

The Different Types of Water Buffalo Breeds

Swamp Buffalo

The swamp water buffalo is mainly found in Southeast Asia and is used as a domestic as well as a draught animal. This breed of water buffaloes is mainly used for the cultivation of rice and its general habitat is marshland. They strive in cool weather conditions and are fitted as pets because of their ability to work. Their eating habits include grass, wood bark, leaves and stems.

River buffalo

The river buffalo is also a common breed of water buffaloes found in India, Hungary, Pakistan, Turkey and Egypt. A river buffalo is very large in size and it also produces milk to a large extent and this is the reason why it can be kept as a pet. These types of buffaloes can thrive in moderate weather conditions and they mostly like to stay in rivers and clean waters. This kind of buffalo usually strives on stem and leaves.

Carabao

The carabao is one breed of water buffaloes that is found in Guam and Philippines. Carabaos are wide, low and heavily built to serve as draught animals and are available in colors of slate gray and light grey. They can survive in hot climate as they have speckled skin and dark mouth and hoofs. Their eating habit comprises of dry hay and wheat germ.

Murrah buffalo

This is one of the widest known buffalo breeds in the world and is generally known for its milk production because of which it serves to be fit as a pet. It is mostly found in Asia, South America and Bulgaria possessing short horns and a massive body. They can dwell in hot and moderate climates where there are sufficient pastures and water.

Romanian buffalo

As the name suggests, the Romanian buffalo is a Mediterranean type of buffalo that is used mainly for milk and this is the main reason why it is reared as a pet. They can swim very easily and their main food usually consists of the grass, the leaves and the stems available in pastures.

Surti

Surti is also a popular breed of water buffaloes that is found mainly in Asia and especially India. They are available in medium size with a long head and convex shape of the horns.

This type of water buffalo is herbivorous and has a very strong condition of living in climates that are humid as well as moderate.

Bhadawari

Bhadawari is also an Asian water buffalo breed that is found mostly in India and is adaptive to terrestrial and tropical habitat. The eating habit of this breed is very similar to the other breeds of buffaloes and it can be said that these buffaloes are tamed as pets because of the high end milk production that they can indulge in.

Niliravi

Niliravi is one exceptional water buffalo breed that is usually very large and comes in blue color. It has a natural habitat that is terrestrial and is also recommended as one of the best water buffalo breeds used as pets. The eating habit of this breed is generally the bark of trees and leaves that are dry along with grass.

Malaysian buffalo

Malaysian buffalo usually originated in Malaysia as the name suggests and can thrive in adverse conditions of the climate. It is found in marshlands and is considered to thrive on food that is green. Stems, leaves, barks and bushes usually make up the eating habit of this buffalo.

Kundi

Kundi is a well-known breed of water buffalo that thrives in cool conditions and is found in Asia. This buffalo is also a draught animal that can be used in the field and this is the reason why it is taken as a pet. The eating habit of Kundi is quite similar to the other buffaloes in the water buffalo category and it also ranks among the various types of popular water buffalo breeds.

Cape Buffalo

The cape buffalo (Syncerus caffer) is sometimes also known as the African buffalo, because it can be found all over sub-Saharan Africa. There are thought to be two subspecies: the savannah type and the forest type. The savannah type weighs between 1,000 and 2,000 pounds and is black or dark brown with closed crescent horns. The forest type has a red-brown coat, is smaller than the savannah type and has horns that curve upward and backward.

The American Buffalo

American buffaloes (Bison bison) measure 5 to 6 feet at the shoulder and can weigh between 800 and 2,000 pounds. As grazers, they mostly live on prairies, plains or in river valleys, but some sometimes be found in forests. They were once populous across North America, but their numbers declined drastically, due to overhunting and loss of habitat, until there were fewer than 1,000 remaining in 1900. Since then, due to careful regulation and breeding programs, their numbers have increased to around 30,000 but they can still only be found in national parks and private rangeland.

Controversy

American buffalo are actually bison. Early settlers and explorers in North America referred to them as buffaloes because they resembled them, but they're quite different. Bison do belong to the Bovidae family, which is the same family both buffalo species belong to, but it's also the family to which domestic cattle and goats belong. Many people believe that it's a misnomer to call an American buffalo a true buffalo.

SELECTION METHODS OF DAIRY ANIMALS

This selection is based on information available on the ancestors like parents, grandparents and great grandparents. The contribution beyond three generations is not much to be considered in **pedigree** selection. Pedigree selection enabled selection at an early age, and selection of males which do not express the traits like milk production through they transmit the genes for the traits.

Individual Selection

Selection is based on the individuals own milk vein, teats, pelvic cavity and udder. This is ideal for characters with high heritabifity. Where as in dairy cattle most of the economic traits have low to moderate heritabilities.

Family Selection

Whale families are selected or rejected as units according to the manphenotypic value of the family. The families may be full sibs or half sibs. The method is useful when the character for which selection is made has' low heritability. Two modifications of family selection applicable to dairy; cattle are sib selection and progeny testing.

Sib Selction

This is a type of a selection where in the selected individuals do not contribute to the family means. This applies to selection of males which do not express the characters and selection of females at an early age.

Progeny Testing

The criteria of selection is the mean value of an individual's progeny which comes close to the breeding value. The value of an individual is judged by the mean value of its progeny known as breeding value. It is equal to the sum of average effects of genes; the individual carries.' Progeny testing pronongs the generation interval. As the bull had to await its progeny test result before 'itsuse, but it is more than made up by the increase in accuracy of selection. A higher intensity of selection is alsd, possible by employing Artificial insemination with pedigree semen.

Economic characters in Dairy Cattle

The various economic characters in Dairy Cattle management are:

- Lactation yield
- Lactation period
- Persistency of yield
- Age at first calving
- Service period
- Dry period
- Inter calving period
- Reproductive efficiency
- Efficiency of feed utilization
- Disease resistance.

Lactation yield

The lactation yield in a lactation period is known as lactation yield. 'The lactation yield in breeds is very low compared to exotic breeds. This is dependent on of calving, frequency of milking, persistency of yield (Normally in dairy cattle 30 - 40 % increase in milk production from first lactation to maturity is observed. After 3 or 4 lactation the production starts declining. For comparison of milk yield of different breeds and animals the mil yield should be converted into fat corrected milk (FCM).

4% FCM = 0.4 total milk + 15 total fat. After parturition the milk yield per day will be increased and reaches peak within 2-4 weeks after calving. This yield is known as peak yield. The maintenance of peak yield for more time is importance for better milk production. The lactation period in breeds is low and so the production is also less and conversion.

Lactation Period

The length of milk producing period after calving is known as lactation period. The optimum lactation period is 305 days. The milk production will be less, if this period is shortened". Breeds will have less lactation period, but in some breeds this period is more with very little milk production

Persistency of Milk Yield

During lactation period the animal reaches maximum milk yield per day within 2-4 weeks which is called peak yield. For high level of lactation yield, this peak yield should be maintained for longer period as far as possible, the maintenance of peak yield for long period is known as persistency, slow decrease in dairy milk yield after reaching peak yield in necessary. High persistency is necessary to maintain high level of milk production.

Age at first calving

The age the animal at first calving is very important for high life time production. Thc dcsirable age at first calving in Indian breeds is 3 years, 2 years in cross breed cattle and 3 1/2 years in Buffaloes. Prolonged age at first calving will have high production in the first lactation) but the life time production will be decreased due to less no of calvings. If the age at first calving is below optimum, the calves born are weak, difficulty in calving and less milk production in first lactation.

Service Period

It is the period between -date of calving and date of successful conception. The optimum service period helps the animal to recover from the stress of calving and also to get back the reproductive organs back to normal for cattle the optimum service period is 60-90 days. If the service period is too prolonged the calving interval prolonged, less no. of calving will be obtained in her life time and ultimately less fife time production.' If the service period is too short, the animal will become weak and persistency of milk production is poor due to immediate pregnancy.

Dry Period

It is the period from the date of drying (stop of milk production) to next calving when the animal is in pregnancy, before next calving. The animal should be given rest period to compensate for growth of focus. A minimum of 2 – 2 ½ months dry period should be allowed) If the dry period is not given or too low dry period, the animals suffer from stress and in next lactation, the milk production drops substantially and also it gives weak calves. On the other hand if the dry period given is too high, it may not have that much effect on increasing milk yield in the next lactation, but it decrease the production in the present lactation.

Intercalving Period

This is the -period between two successive calvings. It is more, profitable to have one calf yearly in cattle and at least one calf for every 15 months in buffaloes. If the calving interval is more, the total no. of carvings in the life time will be decreased and also total life production of milk.

Reproductive Efficiency

The reproductive efficiency means the more number of calves during life time, so that total life time production is increased, the reproduction or breeding efficiency is determined by the combined effect of hereditary and environment several measures of breeding efficiency like number of services per conception, calving interval, and days from first breeding to conception are useful.) Reproductive efficiency has generally a low heritability value indicating that most of the variations in this trait is due to non-genetic factors. In adverse environmental conditions, the poor milk producing animals may not be much effected compared to high effect in high milk yield.

Efficiency of Feed Utilization and Conversion into Milk

The animal should utilize the feed efficiently to convert into the milk.

Disease Resistance

Breeds are more resistant to majority of disease compared to exotic cattle. Cross breeding helps to get this character.

HERD RECORDING

Herd recording (milk recording or herd testing) measures the production of individual cows in the herd. Each cows production is usually recorded at regular intervals of 4 to 8 weeks throughout lactation.

On herd recording (test) day, a sample of milk from each cow is collected to

- measure the volume of milk
- test the fat and protein percentages
- determine the individual cow cell counts (ICCC)

Farmers receive a herd test report after each test day and an annual report which summarizes the performance of the herd for the season.

The herd test report contains:

- Summaries of the best and worst performing cows and highest cell count cows
- Herd trends in production, cell count and comparisons with local averages

- Cow identification, calving date, days in milk, breed, age and sire
- The daily production of milk, fat and protein and the protein and fat test percentage on the day of recording
- The estimated production of milk, protein and fat and the average fat and protein test percentage in the lactation to date
- A production index which enables more accurate comparison between cows in the herd despite differences in age, stage of lactation and time of calving
- A measure of the cows mastitis level as measured by the Individual Cow Cell Count

This report is useful when making decisions on breeding and management of the herd.

How Herd Recording Affects Production and Management

Improving the genetic merit of the herd, through culling low producers and choosing replacements from higher producing dams, has long been considered a key benefit from herd recording. Research and practical farm experience show that major gains come not only from improved genetic merit but also from improved management.

Using the Production Index (PI) to Rank Cows

Many factors affect the level of production of individual cows and make it difficult to determine individual cow performance based on actual production figures.

The production index (PI) adjusts actual production figures to account for age, stage of lactation and time of calving. This process removes the most important factors which cause variation between individual yields.

The PI compares animals within a herd.

Herd Recording and General Management

There is a significant per cow milk production difference between herd recorded and non-herd recorded farms. It is estimated the average per cow production from farms that herd record was more than one third higher than those that do not. However, it is not simply the herd recording causing the increase. Better management decisions made through the use of herd recording data will boost productivity.

Herd recording provides a number of general management benefits. It:

- allows more accurate decisions about which cows to dry off, using individual milk production at the time. In some herds, individual cow cell counts can be used to selectively apply dry-cow therapy and therefore reduce veterinary costs
- allows the use of progeny test (PT) semen. PT semen is cheaper and the average genetic merit of PT bulls should be higher than the average of proven bulls already available
- allows the production level of first-calf heifers to be assessed to see how well replacements are being reared. Poor management of young stock from birth to

calving often results in low production from first-calf heifers. Herd recording is useful to determine if this is a weakness in herd management

- provides information to sell and purchase dairy cattle
- stimulates the interest in the performance of individual cows. Fluctuations in individual production may warn of health problems
- provides electronic data so that the farmer can analyses the herds production records in any way desired. For example, the herds reproductive performance can be closely examined using data collected through herd recording and published in an InCalf Fertility Focus Report

Many herd recording farmers will use farm management software to record further details relating to individual cows. Examples include cases of mastitis, lameness and other disease, mating details, preventative health treatments, workability scoring and calving ease records. This data provides farmers and industry with an increased capacity to analyses herd management contributors to farm productivity.

BREEDING STRATEGIES

Depending on your goals for breeding animals different breeding techniques can be used. The following is a brief overview of some breeding methodologies. The majority are not practical for most people, but for the dedicated animal breeder combinations can be used depending on the breeding strategy.

- Crisscrossing: A continuous program of crossbreeding in which there an alternate use of males is belonging to two breeds. Using two breeds designated as P1 and P2, a crisscrossing program, beginning with the two-breed cross animal (P1 x P2), would begin by backcrossing to one of the parental breeds [(P1 x P2) x P1]. Females resulting from these matings would be bred to a P2 male, [(P1 x P2) x P1] x P2, and so on.
- Genus cross: Mating of animals belonging to different genera (for example, mating of domestic cattle, Bos taurus or Bos indicus, to the buffalo, Bison).
- Grading: Mating of purebred males of a given breed to non-purebred females and the resultant female offspring in successive generations.
- Inbreeding: Mating of animals more closely related to each other than the average relationship within the breed or population concerned.
- Linebreeding: Generally mild form of inbreeding in which animals mated are related to some supposedly outstanding individual.
- Outbreeding: Mating of animals less closely related to each other than the average relationship within the breed or population concerned.
- Species cross: Mating of animals belonging to different species (for example mating of European breed cattle, Bos taurus, to Brahman cattle, Bos indicus).
- Three-breed rotational cross: A continuous program of crossbreeding in which males of three breeds are used on a rotational basis. Using three breeds designated as P1, P2 and P3, the first generation would involve production of two-breed

cross animals, P1 x P5. In the second generation, two-breed cross females would be mated to males of the third breed, (P1 x P2) x P3; three-breed cross females would be mated to males of one of the breeds used to produce the two-breed cross animals, [(P1 x P2) x P3] x P1, and so on.

Descriptions in breeding

- **Progeny = Offspring**
- **Backcross: Progeny resulting from the mating of a two-breed cross animal to one of the parental breed**
- **Crossbred: Progeny resulting from the mating of outcross animals belonging to different breeds.**
- **Inbred line**: Line of animals produced by mating related animals.
- **Incross:** Progeny resulting from the mating of animals from different inbred lines within a breed.
- **Incrossbred: Progeny resulting from the mating of animals from inbred lines of different breeds.**
- **Outcross: Progeny resulting from the mating of unrelated animals within a breed.**
- **Topcross: Progeny resulting from the mating of animals belonging to different families within a breed.**
- **Topcrossbred: Progeny resulting from the mating of inbred males to non-inbred females of another breed.**
- **Topincross: Progeny resulting from the mating of inbred males to non-inbred females of the same breed.**
- **Two-breed cross: Progeny resulting from the mating of males of one breed to females of another breed.**

Table : Table comparisons of results of inbreeding/linebreeding and outbreeding

Trait	Inbreeding/linebreeding	Outbreeding
Uniformity	Good	Poor
Fertility	Poor (to good with selection)	Good
Growth	Poor (to good with selection)	Good
Predictability	Good	Poor
Longevity	Moderate	Good

Breeding Strategies and Plans

Distinct objectives are essential to a breeding plan aiming to improve our livestock breeds for better future profitability such as:

- Higher milk yield
- Better milk composition (depends on your market)
- Good udder or body shape
- Higher beef/meat quality
- Higher growth rate (also related to good nutrition)
- Better disease resistance
- Better fit to available feed and other circumstances
- Color preference
- Polled/non polled etc.

Genes determine the way animals look. For example the coat color in cows is the phenotypic expression of the genes that regulate coat color. The breeder can only plan according to the trait observed. Other genes determine whether the udder will be small or large, whether the animal has horns, the size of the body etc.

Some examples of simple breeding strategies we can mention:

- To increase milk production in zebu cattle: This can be achieved by upgrading the Zebu breed with a proven exotic dairy breed (Jersey, Guernsey, Friesian) upto the fourth generation or the larger and more productive Boran or Sahiwal cow depending on disease pressure in your area.
- To increase disease resistance in exotic dairy breeds: Introduction of animals with better resistance to disease such as Boran or Sahiwal will improve the disease resistance of the dairy breeds. The cross bred offspring may take more after one parent than the other, but there are very good experiences with such crosses.
- Better performance within a breed: If natural breeding is practiced - exchange the bull if the offspring consistently have undesirable traits.

If AI is practiced - look carefully at the pictures and specifications of sires offered to determine whether the proposed sire will add better body conformation to the animal offspring. Some AI semen providers will supply pictures of offspring, with strengths highlighted.

Selection of Breeding Stock

Selection of breeding animals is based on:

- Pedigree or parentage information
- Performance records (Estimated breeding values and phenotype)
- Cost of semen from proven bulls
- Available bulls in the area

CROSSBREEDING SYSTEMS FOR BEEF CATTLE

There are a number of reasons why beef producers could consider crossbreeding rather than traditional straight-breeding programs for their beef enterprises. The first of these is 'grading up' from one breed to another. This has rather limited opportunity other than for those who are establishing new breeds of cattle. The other two reasons are to combine the attributes of two or more breeds to produce an end product that fits market requirements and/or to take advantage of heterosis (hybrid vigor).

Combining the Attributes of Two or More Breeds

The first step is to clearly define the production levels in your current herd for all the economically important traits in your herd. The next step is to determine what production levels are needed for each of those economically important traits, to meet the requirements of the market you are aiming for.

By looking at where you are and where you need to be, you can determine the appropriate breeding objectives for your herd. Identifying breeding objectives is fundamental to planned cattle breeding. Generally, these objectives can be met either by selecting within the current herd or by introducing new breeds into the herd. The introduction of new breeds merely provides a 'quantum leap' in changing the genetics of the herd. However an ongoing selection program is still required to continue to make genetic improvements.

Taking Advantage of Heterosis

Heterosis is the production advantage that can be obtained from crossing breeds, or strains, which are genetically diverse. The new combinations of genetic material can lead to production advantages over and above the average of the two parent breeds or strains. To be of economic advantage, the new production levels need to be above those of either parent strain or breed – otherwise you are better off sticking with the superior parent line.

Heterosis tends to be greater for traits – such as fitness or fertility traits – which are less likely to respond to conventional selection. For example, the level of heterosis achievable for fertility traits is likely to be double that which might be obtained for growth or carcase traits.

Crossbreeding systems

There are five basic crossbreeding systems available to the commercial beef producer.

Two breed cross

The two breed cross system produces first cross, or F1, progeny. In this system, the progeny resulting from the cross of two breeds are usually all sold for slaughter or to another commercial breeder. This system is frequently used in northern New South Wales where there are specific F1 sales.

The system is most useful for situations in which females of a specific breed are well adapted to a given environment. An example is Bos indicus **bred females in north, central and coastal Queensland. These adapted females can be mated to a sire of another breed, resulting in** heterosis for traits such as growth, improved carcase, feed conversion efficiency and vigor.

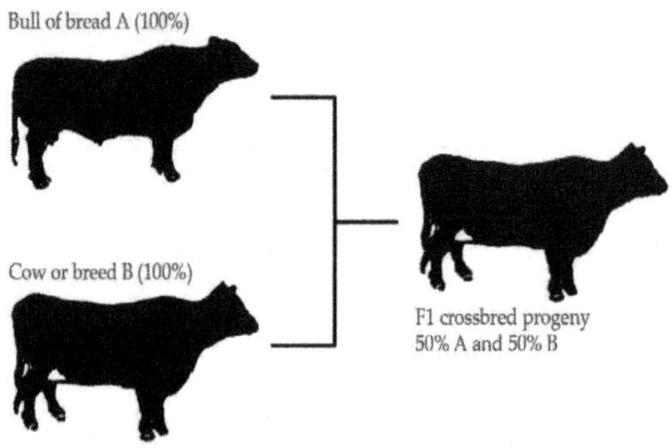

Figure : Two breed cross where breed A and breed B are two purebreds and the F1 progeny (AB) contains equal parts of the two breeds.

Backcross

In a backcross system, all male calves produced from the first cross are sold for slaughter. Female F1 crossbred progeny are mated to males of one of the parental breeds and all progeny are sold for slaughter. This breeding system takes full advantage of heterosis for maternal traits such as fertility of the cow, and milking/mothering ability (as the mother expresses 100% of possible heterosis) and half of the possible heterosis for growth.

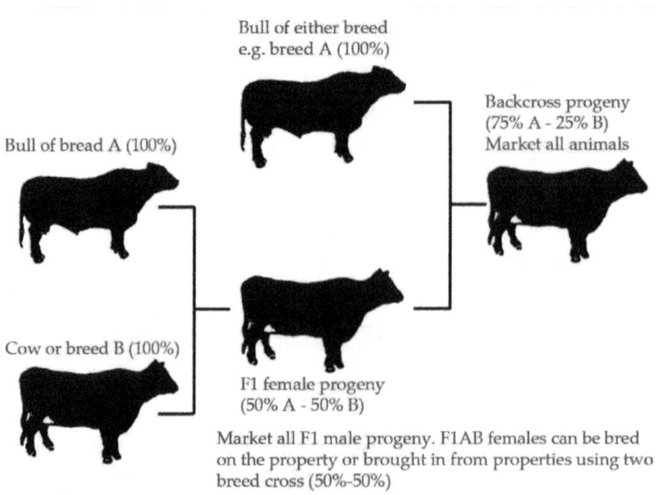

Figure : The backcross is obtained where all the females from a two breed cross are mated to a purebred bull of either of the original breeds. All the backcross progeny are marketed.

This approach is most useful where adaptation to a specific environment is required from a particular maternal breed but where characteristics from the other parental breed are desired for carcase or growth traits. The crossbred F1 female would also have satisfactory environmental adaptation.

Continual backcrossing is the system used by producers to upgrade or change from one breed to another without having to buy purebred cows, such as in development of Brahman herds in Queensland.

Three breed cross

Three breed cross requires the input of three separate breeds. Along similar principles as with backcross, all first cross male progeny are sold. First cross females are joined with bulls of a third unrelated breed, instead of with one of the parent breeds as in the backcross. All progeny of F1 dams are sold for slaughter.

This system takes advantage of both maternal and individual heterosis, and of the complementarity of three breeds. An example would be the a case where the first two breeds are chosen to achieve maternal heterosis and adaptation to an environment, while the third breed (terminal sire breed) produces the most acceptable turnoff animals. This use of the F1 female is generally considered to produce the greatest lift in productivity, but it is influenced by the quality of pure breeds that are maintained to breed the F1 females.

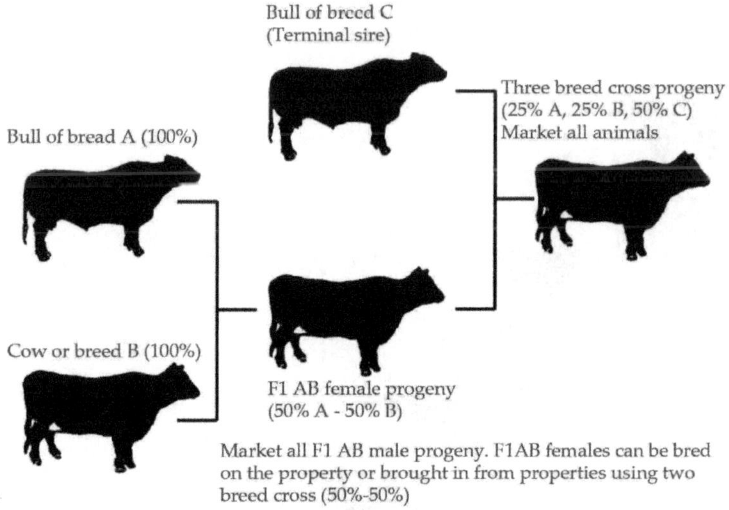

Figure : The three breed cross is obtained when all the females from a two breed cross are mated to a bull of a third, unrelated breed. All the three breed cross progeny are marketed.

Rotational Cross

Rotational crossbreeding, sometimes referred to as sequence breeding, is when males of two or more breeds are mated to crossbred females. Over a number of years, each breed will have contributed its strengths and weaknesses equally.

Variation seen in the progeny in early years of a rotational crossbreeding program may make it more difficult to consistently meet a specific market requirement in this production system. When the breeds used are similar, consistency of performance is less likely to be a problem, although levels of heterosis will also be lower if breeds or breed groupings are relatively closely related.

Levels of heterosis achieved in rotational crossbreeding depend on the number of breeds involved. Once stabilised after many crosses, with a number of breeds (n) contributing equally, the level of retained heterosis may be expressed as:

$(2n-2) / (2n-1)$

Rotational crosses express more heterosis than composites which use the same number of breeds. This increased heterosis in rotational systems is a result of close to maximum heterosis being achieved in each cross with the purebreed.

All animals in the herd benefit from heterosis for both growth and maternal traits in rotational crossbreeding. When three or more breeds are used, heterosis levels of 86% or greater may be achieved. These compare very favorably with self-replacing terminal cross systems in which only about half the progeny show heterosis for growth and only a third of progeny benefit from heterosis in their dam.

In rotational crossbreeding, replacement females are generated for the enterprise from the mating program. All females are potentially available for selection as replacements, whereas in self replacing terminal cross systems, about a third of the heifers are not available for selection. This means a reduced possibility for genetic improvement through heifer selection, placing that pressure on bull selection.

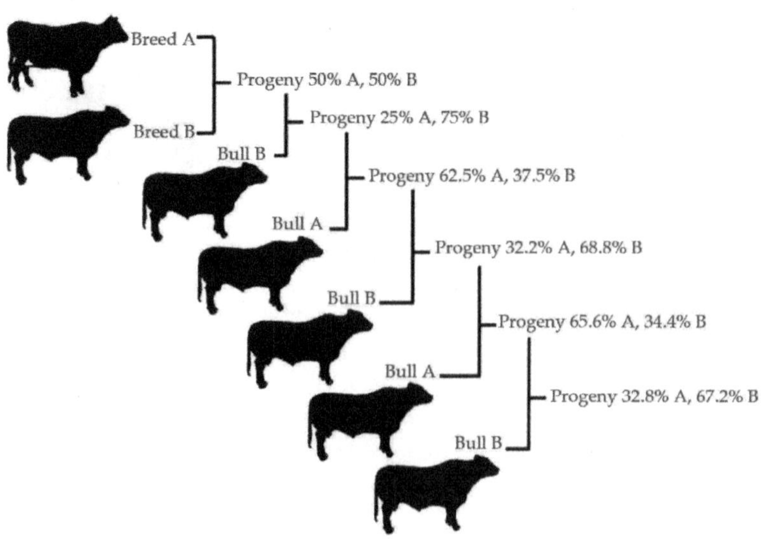

Figure : Starting at 50/50%, the rotation stabilises at 65/35% or 35/65%, giving 65% from the last sire line used.

Because in rotational crossbreeding the mating involves several groups of females of different breed combinations with bulls of different breeds, this system has specific management requirements.

Composite Breed

Development of a composite or synthetic breed results from the crossing of two or more existing breeds. There are many examples of this in Queensland: Santa Gertrudis, Droughtmaster, Braford, Charbray, Brangus and Belmont Red.

The primary advantage of forming composite breeds is that after the initial crosses are made, management requirements are the same as for straight breeding. Should a market signal indicate a change to the characteristics of the composite, there is opportunity to change direction by incorporating another breed or crossbreed.

The percentage of heterosis increases as more breeds contribute in the initial mating program. While the heterosis will not be as high as that achieved with a rotational crossbreeding program with the same number of breeds, the management

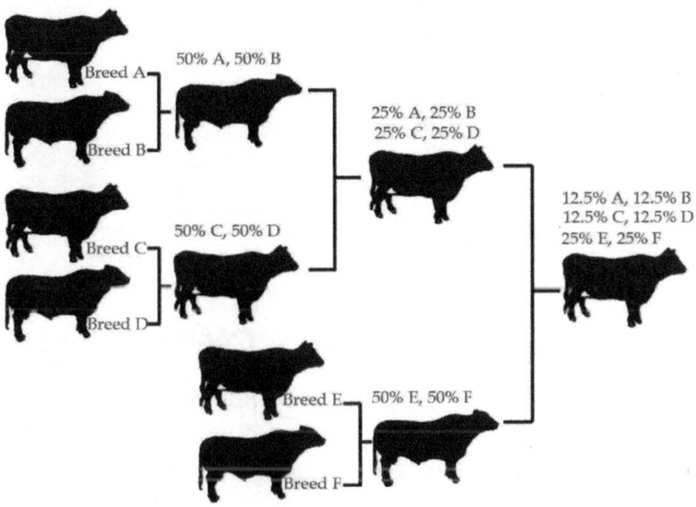

Figure : One simple approach to a Composite breed.

For instance, if there are three breeds in a given composite, the amount of retained heterosis would be expected to be (3-1) ÷ 3 or two-thirds or 67%.

Inbreeding is not usually significant when numbers are greater than 200 or 300 breeders. However, in smaller herds inbreeding can be a problem. In the extreme case where a herd uses only one bull all heterosis is lost after approximately eight generations or 40 years.

Comparing Crossbreeding Systems

The relationship between the various **mating systems**, per cent of maximum heterosis retained and percentage increase in weight of calf weaned per cow exposed is shown in the table below.

Table : Percentage of maximum heterosis expected in progeny

Mating system	Maximum heterosis retained		Superiority over parent breeds	
	Individual	Maternal	Increased weight of calf weight weaned / cow exposed	Increased value of calf weight weaned / cow exposed at $1.30/kg liveweight gain
	(%)	(%)	(%)	(%)
2 breed cross				
A x B	100	0	8.5	16.50
3 breed cross				
(A x B) x C	100	100	23.3	45.00
Rotational crosses*				
2 breed	33	67	12.7	25.00
3 breed	86	86	20.0	40.00
4 breed	93	93	21.7	42.00
Composite crossbreed				
2 breed	50	50	11.6	22.50
3 breed	67	67	15.6	30.50
4 breed	75	75	17.5	34.00
5 breed	80	80	18.6	36.00
6 breed	83	83	19.3	37.50

ARTIFICIAL INSEMINATION (AI) COWS AND HEIFERS

Artificial Insemination (AI) is the second-most common practice of breeding livestock--well, it's the only alternative to breeding livestock next to natural breeding using a male over females. AI is, however, much more common in dairy production than in beef production, though AI is gaining ground in beef breeding herds due to the increased access and marketing of superior and favorable proven sires. Knowing how to AI cattle is important in order to achieve a high success rate in cattle breeding herds where owning a herd bull is not profitable nor recommended.

The following steps are a reasonably detailed article of what is involved in AI'ing cattle. In order to fully understand how AI is done and to even gain certification for being able to AI livestock, visit a bull semen marketing company like Semex, Genex, Select Sires, or any other bull-semen dispersal company. Check to see if they host any programs for AI certification or becoming a certified AI tech that you can attend to learn more about artificial insemination.

Watch the Females, then Get them In

1. Watch your cows and/or heifers for signs of estrus. Females go into heat once every ~21 days, and the heat periods themselves last for 24 hours.

- Check out this article: How to Tell if a Cow or Heifer is in Estrus for physiological, behavioral and physical signs of heat.
    - Most heat periods start or end around dusk or dawn.

2. Twelve (12) hours after estrus is when females should be AI'd. This time period is when the female ovulates, sending an ovum out to the fallopian tube to await fertilization from sperm from a bull.

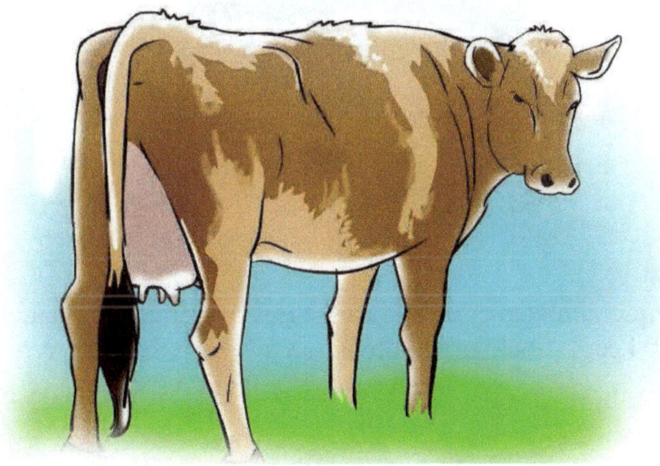

3. Calmly, with proper handling practices put the heifers or cows in to the working alley to the squeeze chute (or alleyway with just a head gate is good enough), and catch the first female in the head gate. If there are others behind her, make sure they are back behind another gate to prevent them from squishing you when they try to move up further in the race. If you have a squeeze chute with a palpation cage attached to it, use this as a means to do your AI'ing. Some barns are set up so that each cow is in a stanchion or head-catch lined up beside one another. This is very convenient for the AI tech who has to AI over 50 cows in one day!

- If doing AI outside, it is best to do it on sunny warm days, not when the weather is rainy or windy or even stormy. If you have a facility that is set up inside a shed or barn, so much the better!

Preparations Prior to Insemination

1. Prepare a water bath that is between 90 to 95 degrees Fahrenheit (or 34 to 35 degrees Celsius) in a thermos. Use a thermometer for better accuracy.

2. Identify which canister contains the semen you need. A semen inventory on the tank that keeps track of the location of each bull prevents any unnecessary searching.

3. Remove the canister from its storage location to the center of the tank. Lift the canister just high enough into the neck region of the tank to grasp the desired cane of semen. Keep the tops of the canister or canes no higher than the frost line, or two to three inches from the top of the tank.

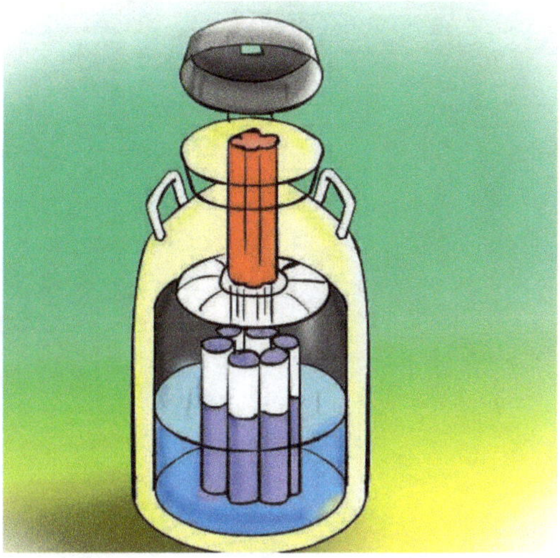

4. Grasp the desired cane then immediately lower the canister to the tank floor. Keep the cane as low inside the tank as possible while removing the straw of semen with tweezers.

- You only have 10 seconds to take the semen straws out!!!

5. Shake the straw to remove excess liquid nitrogen (nitrogen quickly goes into a gaseous state when exposed to air and warmer temperatures)

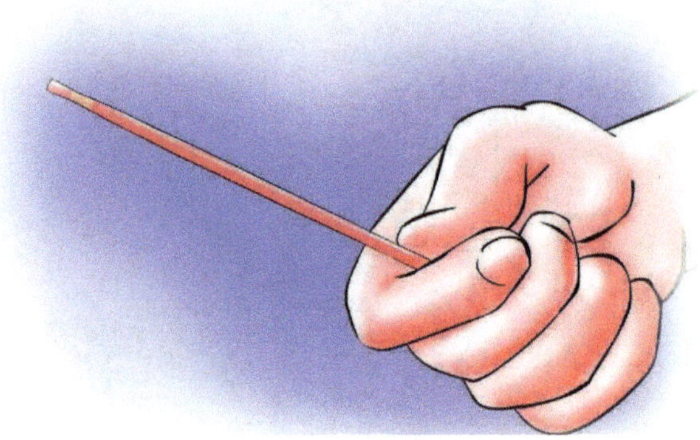

6. Immediately place it the prepared thermos of water and keep it in there for 40 to 45 seconds.

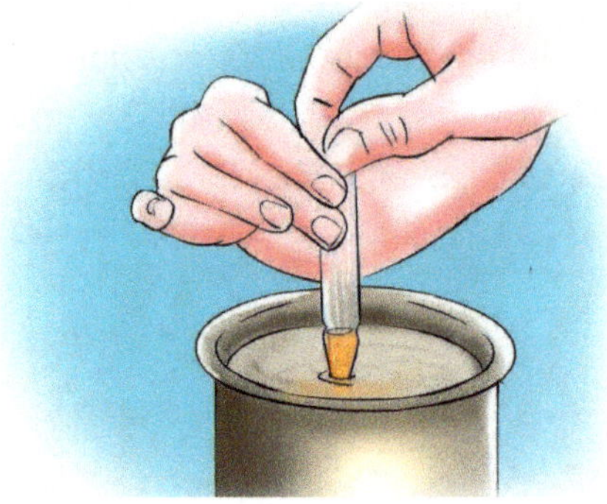

7. After the straw is put in to the warm water, put the can back in the canister by raising the canister up and over the cane and return the canister to its storage position.

- Anytime it takes over 10 seconds to locate a cane the canister must be lowered back into the tank to cool completely. *Never* **return a unit of semen to the tank once you've removed it from the cane.**

8. Get your AI gun ready by having it pre-assembled (this should be done before/ after you prep the thermos with warm water. If it's cold outside, warm the "business end" of the gun up in your coveralls close to your body so it warms up completely. Rubbing a paper towel on the rod also helps warm it up. If it's hot outside, keep it in a cool place. The AI gun should not be extremely hot nor cold to the touch.

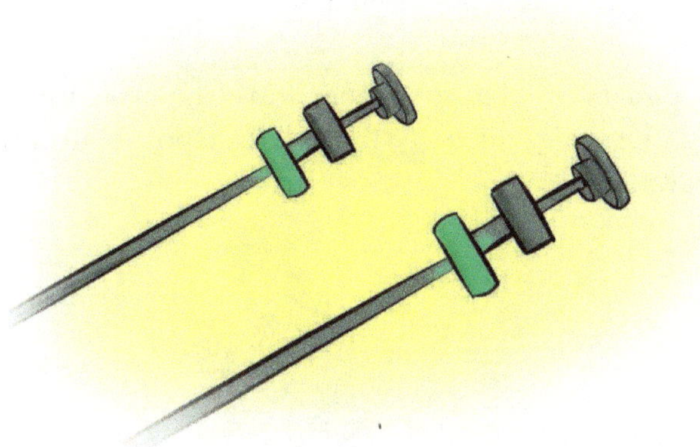

9. Remove the straw from the thermos and wipe it dry with a paper towel. It should be completely dry before you do anything else with it. Flick your wrist slightly, while holding the crimped end, to adjust the air bubble in the straw. The flick should move the bubble to the end you are holding

10. Put the straw in the rod. Clip 1/2" (or 1 cm) from the crimped end of the straw. Use sharp scissors or specially designed straw-cutter scissors should be used, and cut in the area that the air bubble is located.

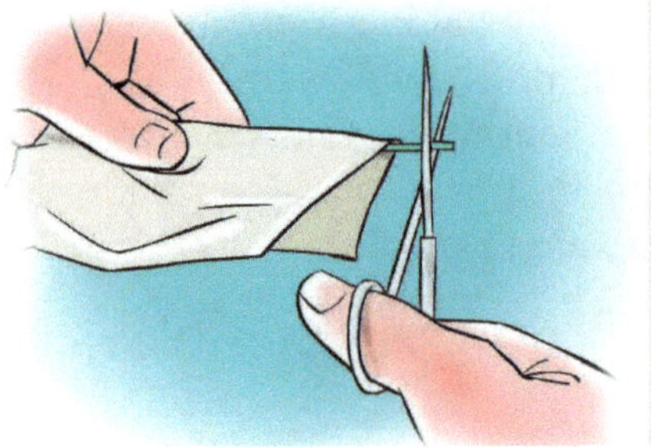

11. Wrap the AI gun in a clean dry paper towel or protective sheath and tuck it in your clothing close to your body to both transport it to the cow and to maintain a constant temperature.

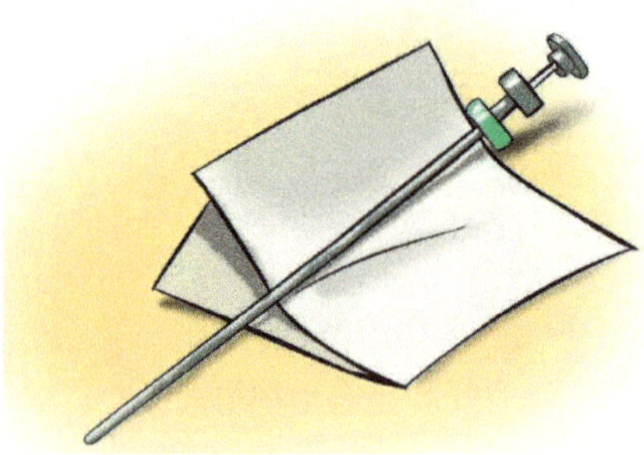

Artificially Inseminating the Female Bovine

1. Move the tail so it's on top of your left forearm or tie it up so it will not interfere with the AI process. Raise the tail with one hand (preferably the right) and with the other (which should be gloved and lubricated), gently reach inside the cow to clean out any feces that may interfere with the process of feeling for and inserting the AI gun into the cow's vagina.

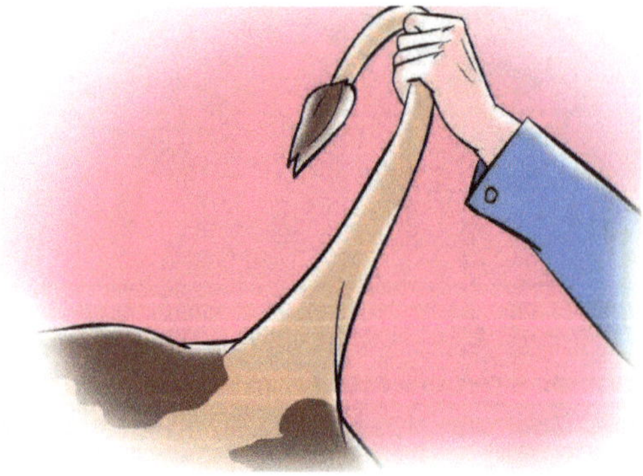

5. Clean the vulva with a clean paper towel or rag to remove excess manure and debris.

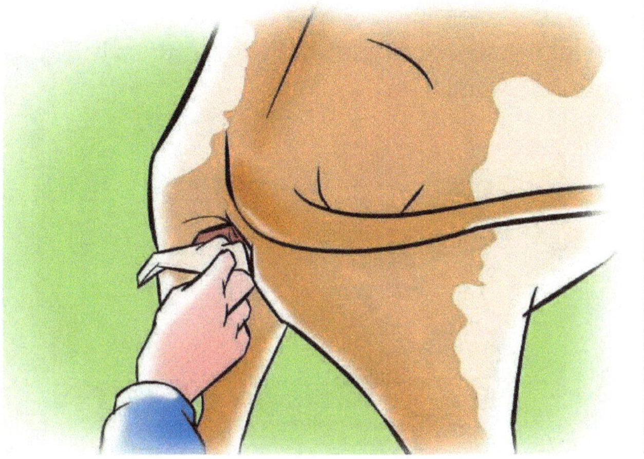

3. Take the gun out of your jacket or overalls, unwrap it, then insert it at a 30 degree angle into the cow's vulva. This is so that you avoid going into the urethral opening into the bladder.

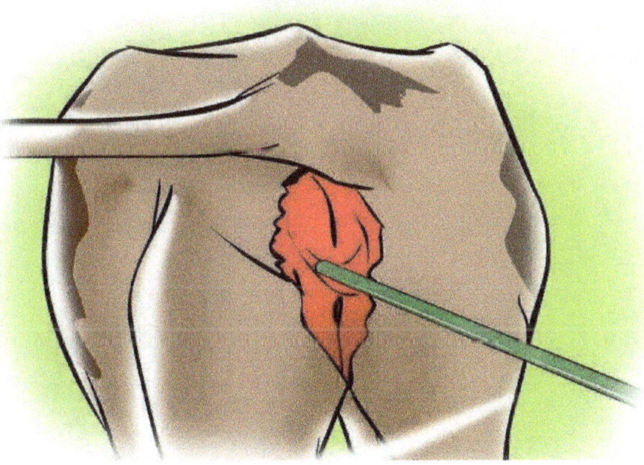

4. With your left hand in the rectum of the cow (which should have been there to begin with), feel with your fingertips through the wall of the rectum and vagina the location of the end of the AI gun until you reach the cervix.

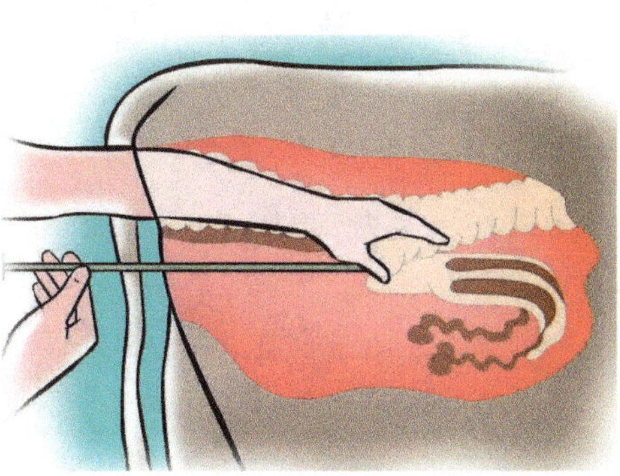

5. Grasp the cervix with the hand in the rectum of the cow (like you would hold a bar that is below your hand) and hold it steady while you thread the rod into and through the cow's cervix.

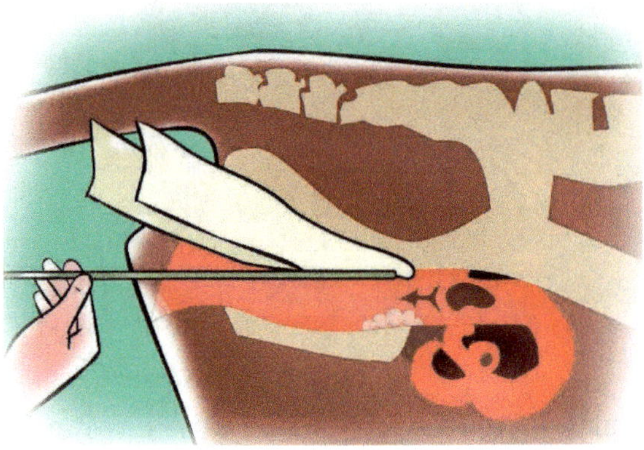

6. When the rod is all the way through the cervix, check the location with your index finger. **The rod should be only 1/2 to 1/4 of an inch into the uterus.**

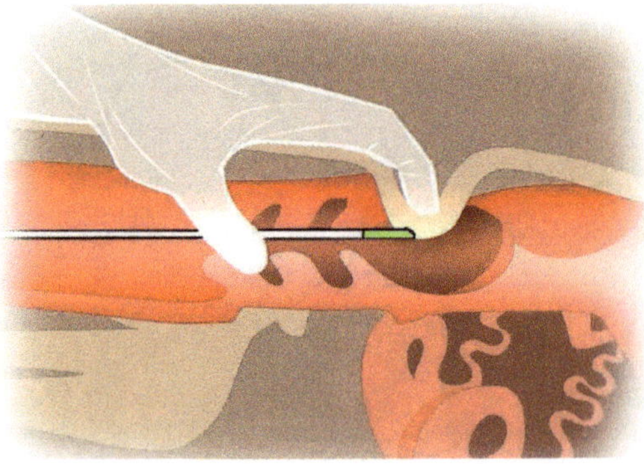

7. Slowly depress the plunger at the end where your right hand is so that 1/2 or is deposited.

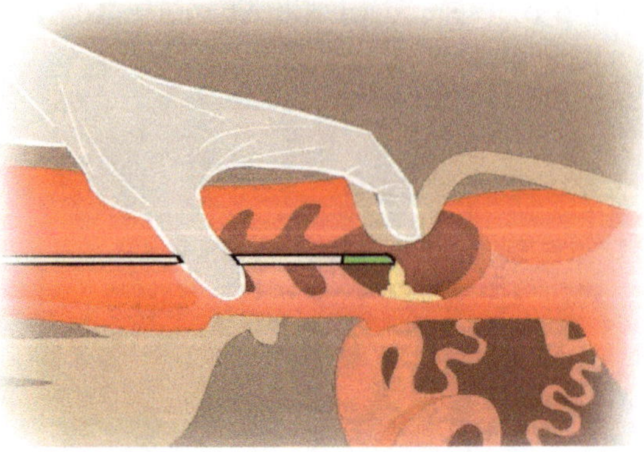

8. Recheck the location of the semen to make sure you are in the cow's uterus and not in any of her "blind spots", and deposit the other half of the straw's contents.

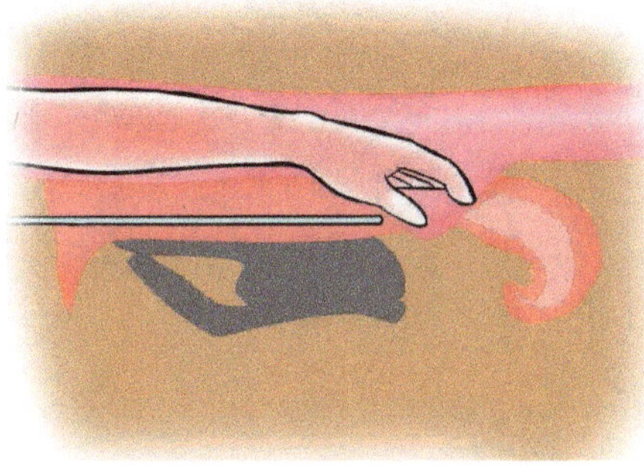

9. Slowly remove the AI gun, your hand and arm from inside the cow. Check for any blood, infection or semen "feedback" from inside the sheath.

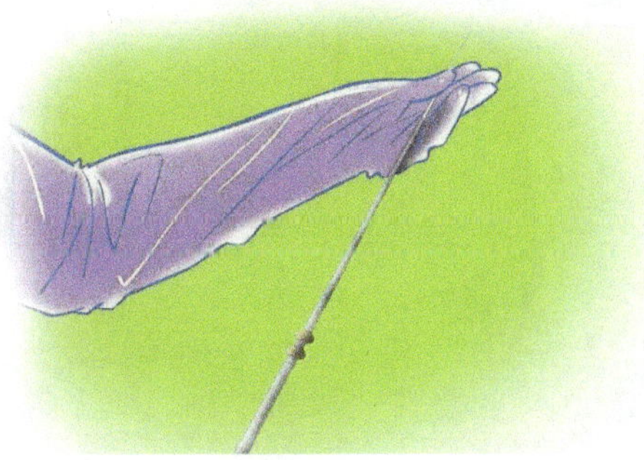

10. Recheck the straw to see if you used the right bull semen for the cow.

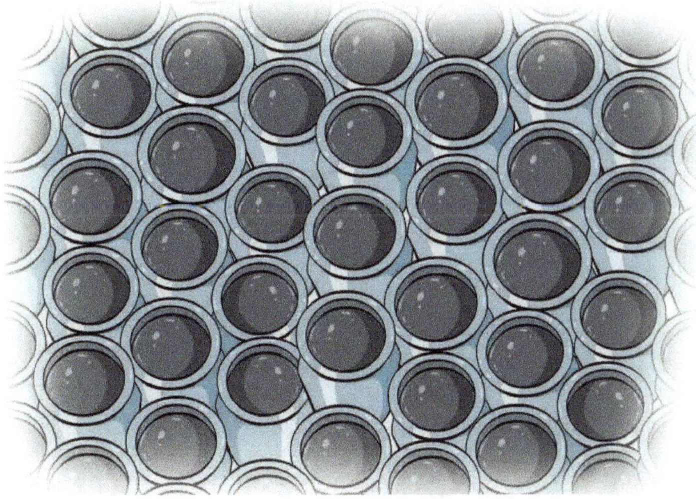

11. Dispose the straw, glove, and towels in the proper place.

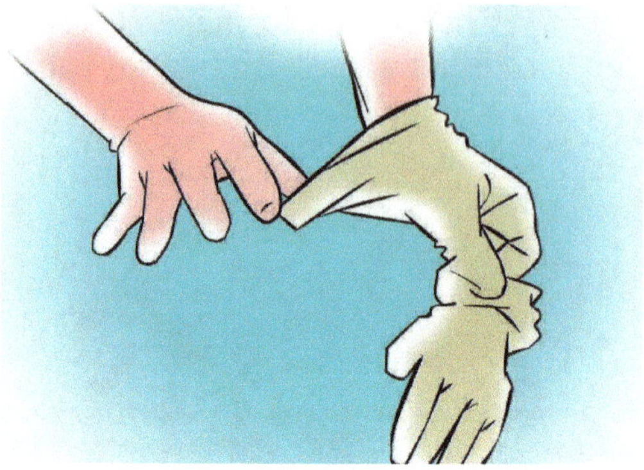

12. Clean the AI gun if necessary.

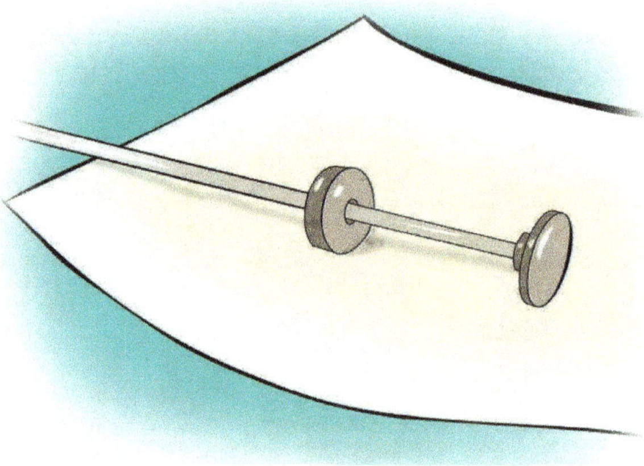

13. Record breeding information on any record keeping system you have on hand.

14. Release the cow (if necessary, depending on the breeding set-up you have) and restrain the next one to be inseminated.

15. Recheck the temperature of the water in the thermos before repeating the steps above for another cow.

16. Repeat the steps above for the next cow.

REFERENCES

1. Animal Care and Use Committees Bibliography. T. Allen and K. Clingerman. 1995. Beltsville. Md.: U.S. Department of Agriculture, National Agricultural Library (Publication #5RB92-16). 38 pp.
2. Animal Care and Use: Policy Issues in the 1990's. National Institutes of Health/ Office for the Protection from Research Risks (NIH/OPRR). 1989. Proceedings of NIH/OPRR Conference, Bethesda, Md.
3. Cost Analysis and Rate Setting Manual for Animal Resource Facilities. Animal Resources Program (ARP), Division of Research Resources (DRR), National Institutes of Health (NIH). 1979 revised. NIH Pub. No.80-2006. Washington, D.C.: U.S. Department of Health, Education and Welfare. 115 pp. (Available from ARP, DRR, NIH. Building 31, Room 5B59, Bethesda, MD 20205).
4. Effective Animal Care and Use Committees. F. B. Orlans, R. C. Simmonds, and W. J. Dodds, eds. 1987. In Laboratory Animal Science. Special Issue, January 1987. Published in collaboration with the Scientists Center for Animal Welfare.
5. Essentials for Animal Research: A Primer for Research Personnel. B. T. Bennett, M. J. Brown, and J. C. Schofield. 1994. Beltsville, Md.: National Agricultural Library. 126 pp.
6. Guide to the Care and Use of Experimental Animals, Volume 1, 2nd ed. E. D. Olfert, B. M. Cross, and A. A. Mc William, eds. 1993. Ontario, Canada: Canadian Council on Animal Care. 211 pp.
7. Institutional Animal Care and Use Committee Guidebook. NIHIOPRR. 1995. NIH. Pub. 92-3415. (IACUC duties. special considerations. federal regulations. references and resources.)
8. Laboratory Animal Medical Subject Headings, A Report. NRC (National Research Council). 1975. A report of the ILAR (Institute of Laboratory Animal Resources) Committee on Laboratory Animal Literature. Washington, D.C.: National Academy of Sciences. 212 pp.
9. Reference Materials for Members of Animal Care and Use Committees. D. J. Berry. 1991. Beltsville, Md.: U.S. Department of Agriculture, National Agricultural Library (AWIC series #10). 42 pp.

CHAPTER 6

Basics of Heat Detection and Artificial Insemination

Chapter Contents

- **Heat Detection and Timing of Insemination for Cattle**
- **Characteristics of the Estrous Cycle**
- **Signs of Heat**
- **Factors Affecting Estrous Behavior**
- **Estrous Detection Program**
- **Timing of Insemination**

A successful AI program must include efficient and accurate heat detection and timely AI relative to ovulation. The failure to detect heat is the most common and costly problem of AI programs and the major limiting factor of reproductive performance on many dairies (Nebel and Jobst, 1998). This is evidenced by the reality that heat detection efficiency (defined as the proportion of possible heat periods detected during a period of time) is less than 50% in many herds. Nebel (2003) argues that skilled managers achieve 65 to 75% heat detection efficiency. Heat detection accuracy (defined as the proportion of detected heat periods in which cows exhibited low progesterone concentration in milk or blood) varies widely, as up to 60% of cows presented for AI did not exhibit low progesterone at the time of AI.

HEAT DETECTION AND TIMING OF INSEMINATION FOR CATTLE

Efficient and profitable reproductive performance of a dairy herd requires routine but conscientious heat detection and proper timing of artificial insemination. Failure to detect estrus (heat) is a major factor contributing to low fertility. Approximately half of the heats are undetected on dairy farms in the United States. In addition, research based on levels of the hormone progesterone in milk shows that up to 15 percent of the cattle presented for insemination are not in heat. Failure to detect cows that are in heat and breeding cows not in heat result in economic loss for the producer because of extended calving intervals and additional semen expense. If a herd producing 16,000 pounds of milk per year and maintaining a 12.5 month calving interval is compared with the same herd maintaining a 13.5 month calving interval, the loss in potential milk alone is between $35 and $45 per cow per year, depending on milk price and feed costs. Therefore, poor heat detection is costly to the producer and should be considered the critical component of reproductive management.

Efficient heat detection and timely insemination also are important to beef producers who use artificial insemination. Failure to detect estrus early in the breeding season or improper timing of insemination due to heat detection errors results in extended calving intervals and additional semen expense.

This publication provides information on characteristics of the estrous cycle, signs of heat and estrous behavior, factors affecting expression of estrus, and management suggestions for improving heat detection. Various ways to monitor the efficiency of heat detection and several heat detection aids also are described.

CHARACTERISTICS OF THE ESTROUS CYCLE

It is important for those who manage cattle to understand the changes in ovarian structures and hormone concentrations during the normal estrous cycle.

In general, there are two structures that can be found on the ovary, the corpus luteum and follicles of various sizes. It was thought that several follicles developed large fluid cavities (antrum) toward the end of the cycle or prior to the first postpartum heat but that only one or two follicles were dominant and ovulated, releasing the oocyte (egg) shortly after heat. The other follicles were thought to degenerate and a corpus luteum to develop at the site of ovulation. Recent studies, however, reveal different changes in growth and degeneration of follicles during the cycle.

Using real-time ultrasound technology to monitor ovarian structures on a daily basis, several independent research groups have shown that cows or heifers may have two, three, or four groups of follicles that develop during each estrous cycle. Usually, a single follicle within each group becomes dominant, suppressing the continued growth of the other follicles within that group. Such groups of developing follicles are called "waves" of follicular growth. Thus various populations of small, medium, and large follicles are present on the ovary each day of the cycle.

On about day 18 of a normal 21-day cycle, the corpus luteum (CL) that developed after the previous ovulation begins to regress and progesterone concentrations decline. The dominant follicle of this last wave continues to increase in size and produces the hormone estrogen. Since progesterone concentrations are low, estrogen causes the characteristic signs and behavior associated with estrus. At the beginning of standing heat, estrogen also initiates the release of a surge of luteinizing hormone (LH) from the anterior pituitary gland in the brain.

Luteinizing hormone begins the process of ovulation, which occurs approximately 25 to 32 hours after the onset of standing heat. At the site of ovulation, the new CL increases in size and gradually produces increasing amounts of progesterone. Concentrations of progesterone in the blood remain high from about day 6 through day 18 of the cycle. Although follicles continue to develop and even produce estrogen (estrogen-active follicles) during midcycle, the high level of progesterone prevents their final maturation and inhibits expression of heat. Eventually these larger follicles degenerate and estrogen production declines.

Progesterone also prepares the uterus for pregnancy, inhibits uterine contractions, and maintains pregnancy. The CL regresses in cows that fail to become pregnant and progesterone levels decline, the dominant follicle completes development, and the cow returns to estrus, initiating a new cycle. If pregnancy occurs, the CL does not regress and progesterone levels remain elevated throughout pregnancy.

Thus the estrous cycle can be divided into four periods: proestrus, estrus, metestrus, and diestrus. Proestrus is the period when progesterone declines with regression of the CL, estrogen increases, and secondary signs of estrus begin to occur. Estrus is characterized by standing behavior (true heat). Metestrus begins immediately after estrus when ovulation and early development of the CL occur, a period that lasts three to five days. Finally, diestrus is the time when the CL is functional, the longest phase of the estrous cycle.

Figure 1 illustrates the structural changes on the ovaries and the hormone concentrations during an estrous cycle with three waves of follicular development. Although specific days of the cycle are indicated on the diagram, this is for illustration purposes only since there is variation among cows.

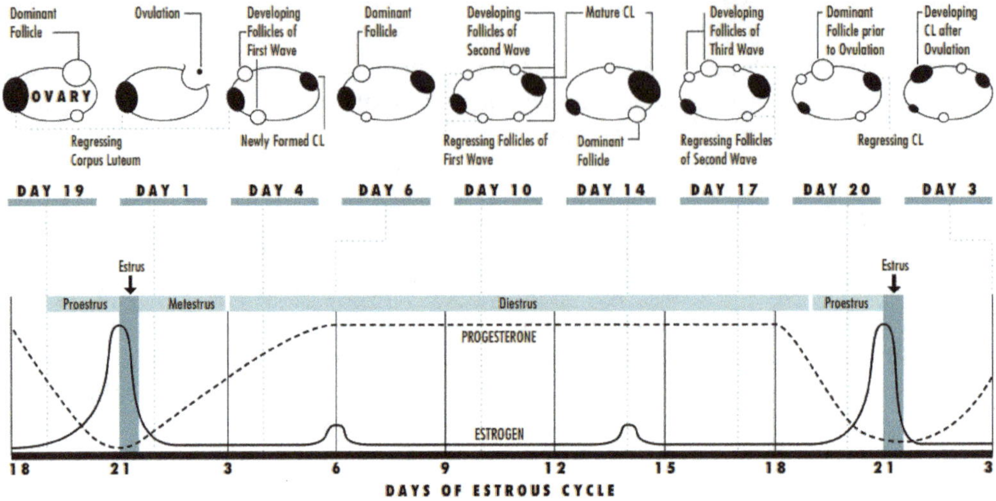

Figure 1. Changes in ovarian structures and hormone concentrations during the estrous cycle. For simplicity, only one ovary is illustrated; however, structural changes are occurring on both ovaries. Specific days of the cycle are listed for illustration only. There is variation among cows.

SIGNS OF HEAT

It is essential to understand the primary and secondary signs of heat in order to achieve accurate and efficient heat detection.

Standing and mounting activity.

Primary sign

A cow standing to be mounted is the most accurate sign of estrus. *Standing heat* is the most sexually intensive period of the estrous cycle. During this period, cows stand to be mounted by other cows or move forward slightly with the weight of the mounting cow. Cows that move away quickly when a mount is attempted are not in true estrus. In order for standing behavior to be expressed, cattle obviously must be allowed to interact.

The expression of heat is due to the elevated level of estrogen in the blood when progesterone is very low. Occasionally cows in early pregnancy, approaching the end of pregnancy, or with ovarian follicular cysts have similar hormonal relationships and may express signs of heat.

The average duration of standing heat is 15 to 18 hours, but heat duration may vary from 8 to 30 hours among cows. An estrous cow usually stands to be mounted 20 to 55 times during her estrous period. Each mount lasts three to seven seconds. Factors that affect expression of heat are discussed in a later section.

Secondary signs

Secondary signs vary in duration and intensity. These signs may occur before, during, or after standing heat and are not related to time of ovulation. Dairy producers should use these signs as clues or watch the specific cow more closely for standing behavior. To determine whether cows exhibiting secondary signs will stand to be mounted, they may be isolated with a sexually active cow or teaser animal.

Mounting Other Cows

Cattle that exhibit this behavior may be in heat or approaching heat. Mounting activity is performed much less frequently by cows in midcycle. Although mounting cannot be used as a true primary sign of heat, cows exhibiting such behavior should be watched closely for standing behavior.

Mucus Discharge

As an indirect result of elevated estrogen levels, mucus is produced in the cervix and accumulates with other fluids in the vagina before, during, and shortly after estrus. Long viscous, clear elastic strands of mucus generally hang from the vulva. Sometimes, however, the mucus does not appear externally until the cow is palpated during insemination and the mucus is expelled. Mucus also may be smeared on the tail, thighs, flanks, or perineal region.

String of clear mucus being discharged.

Swelling and Reddening of the Vulva

During heat the vulva swells and the interior becomes moist and red. Generally these symptoms appear before heat and remain for a short period after heat. Thus they are not precise indicators of estrus. During midcycle the lips of the vulva are pale and more difficult to separate.

Bellowing, Restlessness, and Trailing

Cows in heat are more restless and alert to their surroundings. When allowed to interact with other cattle, cows coming into heat "proestrus" and cows in heat persistently trail behind to try to mount other cows. Research shows that cows in heat spend less time resting than nonestrous herdmates. Prior to and during heat, they remain standing and alert while their herdmates are lying down and resting. This is more noticeable for stanchioned cows. Cattle may bellow more frequently during estrus. Although these are not definitive signs of heat, cows exhibiting such behavior should be watched closely for standing behavior.

Rubbed Tailhead Hair and Dirty Flanks

As a consequence of being ridden, the hair on the tailhead and rump is fluffed-up, rubbed, or matted, and the skin may be exposed. The legs and flanks may be smeared with mud or manure.

Chin Resting and Back Rubbing

Prior to mounting, cows often rest or rub their chin on the rump or back of the cow to be mounted. This application of pressure may be considered a test for receptivity to being mounted. Both cows should be observed closely for mounting and standing behavior.

Chin resting and trailing activity.

- Sniffing Genitalia

Sniffing the genitalia and licking the vulva of other cows occur much more frequently with cows in proestrus and estrus.

Head Raising and Lip Curling

Generally this activity follows sniffing of the genitalia and occurs more frequently if the cow being investigated is in heat and urinates.

Decreased Feed Intake and Milk Yield

Estrous cows spend less time feeding. Some studies also have reported decreased milk yield during estrus, and a Canadian study reported a slight increase in milk yield toward the end of heat. But since many factors other than estrus can affect milk yield on a specific day, this is not a reliable indicator of estrus.

Metestrous Bleeding

Some cows and most heifers have a bloody mucus discharge one to three days after estrus, but onset of this symptom, called metestrous bleeding, is quite variable. High estrogen levels during estrus cause blood to leak from vessels near the surface of the uterus. This discharge indicates that the cow was in heat and does not mean that she failed to conceive. However, such animals should be watched closely for a return to heat in 18 or 19 days.

FACTORS AFFECTING ESTROUS BEHAVIOR

Various factors related to environment, cow health and nutrition, and herdmates can affect estrous behavior.

Arrangement where cattle are crowded.

Type of Housing

Any housing arrangement that allows cattle to interact throughout the day provides more opportunity for mounting and standing behavior to be expressed. Cattle housed in tie-stall or stanchion barns must be turned out in order for this behavior to be expressed. If cattle are not allowed to interact, the herd manager must use less reliable secondary signs to determine which cows are in heat. Caution: It would be incorrect to assume that more heats are observed in loose-housing or free-stall barns than in tie-stall or stanchion barns. Although there is a greater opportunity to observe heat when cows interact frequently in free-stall barns, taking time to observe estrous behavior is still necessary.

Footing Surface

To what extent does a slippery footing surface affect expression of estrus? Research conducted in North Carolina compared estrous activity of high-producing Holstein cows that were watched for heat for one hour every eight hours, 30 minutes on dirt and 30 minutes on grooved concrete. Table 1 shows total mounts and stands during estrus and average number of mounts and stands during a 30-minute heat check period. Duration of heat was longer for cows observed on dirt. Mounting and standing behavior were nearly doubled when cows were checked for heat on dirt as compared to concrete.

Table 1. Mounting and standing activity of thirteen Holstein cows on dirt and concrete

Activity	Dirt	Concrete
Source: J. Britt et al. (1986), Journal of Dairy Science 69: 2195.		
Duration of heat (hours)	13.8	9.4
Total during estrus		
Mounts	7.0	3.2
Stands	6.3	2.9
Average during 30-minute heat check		
Mounts	3.7	2.5
Stands	3.8	2.7

A second study also confirmed that mounting activity occurs more frequently when cows are on dirt rather than concrete. When five estrous cows were individually presented with an opportunity to spend time on dirt or on concrete in the presence of a tied cow which was either an estrous cow or a cow not in heat, the test cow spent an average 70 percent of the time on dirt. The test estrous cow mounted more frequently when a tied estrous cow was on dirt rather than on concrete.

Finally, in a Purdue University study in which heifers were housed in a free-stall barn with access to a bedded pack and observed continuously for estrous behavior over a 96-hour period, 70 percent of the mounting activity occurred in the bedded pack area. Less activity was observed in the dry lot, in the free-stall area, and near the feed bunk (Table 2).

Table 2: Percentage of mounting and standing to be mounted behavior in various locations in a loose-housing arrangement

Location	Mounts and standing behavior	
	Number	% of total
Source: J. Pennington et al. (1985), Agri Practice, Vol. 6, No. 9.		
Bedded pack	1098	70
Dry lot	204	13
Feed bunker of manger area	108	7
Free-stall area	94	6
Free-stalls	55	4

Icy or slick concrete is always a problem for free-stall herds. Concrete should be grooved or scabbled to provide traction. Moving cattle to a dirt lot or from one area to another not only provides a better surface but also provides added stimulation from the movement alone.

Feet and Leg Problems

Cows with sore feet or legs or poor structural conformation exhibit less mounting activity, or they stand to be mounted when not in heat because it is too painful to avoid being mounted. One British study involving 770 cows with nearly 1500 lactations showed that lameness caused by specific lesions on the hoof was associated with a 7-day increase in days to first service and 11 more days open compared to herdmates without lameness. These differences were greater for cows with sole lesions that developed between 36 and 70 days postpartum, the time when cows should first be detected in heat. For those cows the interval to first service and days open increased 17 and 30 days, respectively.

Cow Density

Data are not available for determining optimal number of open cows per unit area or total number of cows per pen to achieve maximum expression of estrous behavior. However, common sense would suggest that if cattle are forced into a crowded area, such as a holding pen associated with a milking parlor, that mounting activity would be suppressed or cows would be forced to stand to be mounted even when they are not in heat. Too often the only time cows are observed for heat is when they are in holding pens prior to milking. In overcrowded free-stall barns, cows may stand to be mounted in a stall or alley simply because they cannot escape a mounting cow. Crowded conditions also may increase the incidence of false-positive heats determined by rump-mounted heat detection devices or markings.

Temperature

Research conducted in Virginia showed that as maximum daily temperatures increased to about 75°F, mounting activity also increased. But at temperatures above 85°F, beyond the comfort range for cattle, mounting was less frequent. Research done at Purdue University suggests that cows in cold weather have more mounting activity than cows exposed to hot weather; however, during hot weather estrous cows tend to exhibit more secondary signs, such as rubbing, licking, and chin resting.

Variation During the Day

Do cows prefer to mount other cows at a certain time of day? Herd managers often remark that most mounting occurs in early morning or during the later evening hours. In a study conducted in Canada, video cameras monitored estrous activity in a free-stall area 24 hours a day. This study showed that nearly 70 percent of the mounting occurred between 7 p.m. and 7 a.m. (Figure 2). This observation suggests that cows are most likely to exhibit mounting activity when they are not distracted by farm activities such as feeding, milking, and barn cleaning. The data may further indicate that cattle prefer to mount during the cooler times of day.

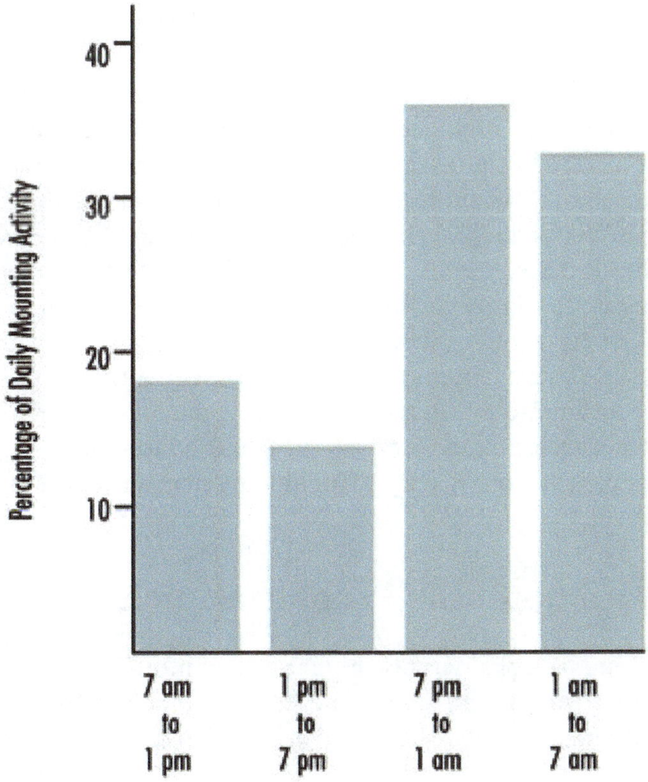

Figure 2. Effect of time of day on mounting activity for estrous cows. Adapted from Hurnik et al. (1975)

Herdmate Status (Stage of Estrous Cycle)

To determine the cause of silent heats or poor expression of heat, consider the overall reproductive status of the herd and compare the ratio of open to pregnant animals in each group of cows and heifers. Pregnant cows are the least likely group of herdmates to mount a cow in heat. Herd managers must rely on the other open cycling herdmates to detect heat. What about the other nonpregnant herdmates? Do herdmates vary in their ability to detect and mount cows in heat? Does the day of the herdmate's estrous cycle influence her interest in the estrous cow?

When Penn State researchers conducted a very controlled study, estrous cows encountered each of 19 cycling herdmates in an isolated barn one-to-one for 10 minutes. The total number of mounts by each herdmate was determined on day 10, day 15, day of heat, and day 5 of the next cycle. Herdmates varied in mounting activity, as did estrous cows in attracting mounts. More important, for the herdmates as a group, the number of mounts differed substantially with the day of the cycle. Herdmates in midcycle (day 10 and day 15) and on day 5 averaged one mount or less during the 10-minute observation period. Some herdmates did not even mount the estrous cow on those days. However, when the herdmate herself was in estrus, the number of mounts averaged 2.5 during the 10-minute period. Thus, taken as a group, cows are poor heat detectors in the middle of their cycle. Since this represents 50 to 60 percent of the estrous cycle and reduces the number of effective heat-detecting herdmates by a similar percentage, managers must rely on cows in or near heat to detect other estrous cows.

In some herd situations, a few open cycling cows may contribute to the problem of silent or missed heats. In small herds, most of the herd may be pregnant at certain times, and the stage of the cycle of the few nonpregnant cows may be such that they are not effective heat detectors. As more animals become pregnant, the number of potential heat-detecting animals is reduced. The situation is similar for herds that freshen on a seasonal basis. After an intensive breeding period, when a high percentage of the herd is pregnant, it becomes increasingly difficult to identify the few open cycling cows in heat. There simply may not be enough herdmates in the proper stage of the cycle to interact with an estrous cow. A third situation may occur frequently in free-stall herds where cows are grouped according to production. Generally, the lower production group contains pregnant cows, but because of low production, some nonpregnant cycling cows may be included in this group. It becomes very difficult to detect the cows in heat because their herdmates are pregnant.

Number of Herdmates in Proestrus or Estrus

Frequency of mounting is considerably higher when more than one cow is in heat or approaching heat (proestrus) at the same time. Canadian scientists found that the number of mounts increased significantly when two or more cows were in heat simultaneously (Table 3).

Table 3. Effect of number of cows in heat on mounting activity.

Number of cows	Average mounts per cow in heat
Source: J. Hurnick et al.(1975), Applied Animal Ethology 2:55.	
1	11.2
2	36.6
3	52.6
4 or more	49.8

Nutritional Factors

Mounting activity and other sexual behavior have been shown to decrease in cows that lost more weight since calving than herdmates with minimal weight loss. On the other hand, there are certain cows within the herd or entire herds in which the majority of the cows are truly anestrus (noncycling). Possible causes of anestrus include poor body condition, anemia, uterine infection, cystic ovaries, and parasitism.

Lactation Number, Days Postpartum, and Milk Production

With the possible exception of heifers, lactation number does not seem to affect estrous behavior. Variable results have been obtained from research comparing estrous behavior with level of milk production or number of days postpartum. These factors have much less influence on the expression of estrus than other factors described above.

ESTROUS DETECTION PROGRAM

The following management practices may aid in estrous detection.

Heifers on dirt lot with good footing.

- Use your time efficiently, observing for heat when cattle are likely to mount.
- Allow cows to interact, especially during the evening and early morning hours, when most of the mounting activity occurs. Even though loose housing systems provide more time for cow interaction, be sure to observe the cattle frequently. Move pastured cattle to an area where they easily can be observed. In conventional housing systems, turn cows out twice daily for 20 to 30 minutes. Be sure to turn cows out when time can be spent observing them. Avoid scheduling observation periods at feeding time or during the warmest hours in summer.

An aggressive heat detection program can be effective. A Pennsylvania study involving approximately 200 repeat breeding heifers (three or more services) compared the average and distribution of cycle length based on farm records before the heifers arrived at the research station with these estrous cycle characteristics after arrival. The major difference in management was that the heifers were turned out for heat detection at 8:00 a.m., 4:00 p.m., and midnight. The average duration of cycle length and the frequency of long cycles were reduced. The onset of heat occurred equally for the three heat detection periods. Table 4 shows that the more frequent the observation, the more heats are detected.

- Slippery and muddy conditions severely inhibit mounting activity. Provide an area with a good footing surface where cattle are free to interact and where few obstacles hinder movement. Moving cattle to a separate area for heat detection may stimulate estrous behavior.
- When cows have sore feet and legs, heat detection is more difficult. Minimize this problem by trimming hoofs periodically, and treat infected feet as soon as a problem is apparent.

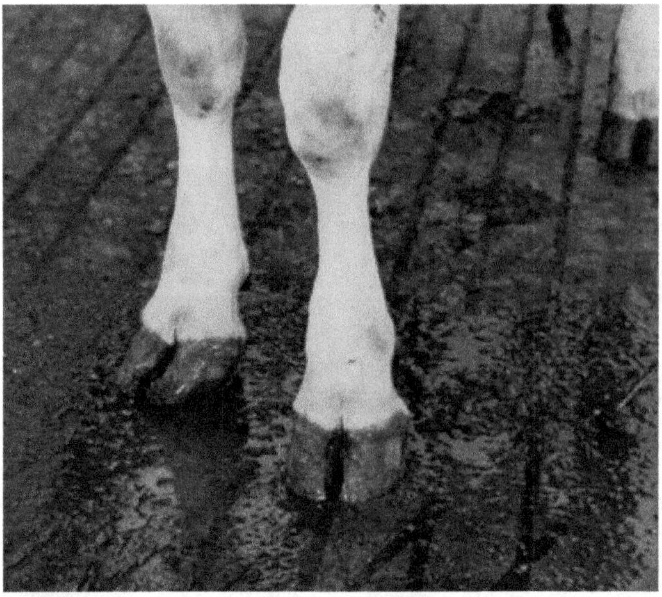

Cows on grooved concrete.

When several people are working with the herd, assign one person to be responsible for heat detection, and allow time for employees to do the job properly. Train employees to recognize signs of heat and promptly report this information to the responsible person. Consider having a financial incentive program to increase heat detection efficiency.

- Studies have shown that up to 15 percent of the cattle presented for insemination are really not in heat. Poor cow identification can be one cause of this problem. Legible neck chain numbers, large ear tags, and freeze brands can aid in accurate identification and can reduce mistakes.
- Record all heats, whether the animal is to be inseminated or not. Heat detection will improve if future heats can be anticipated. Use a pocket notebook to record heats and other information. Transfer information to a heat expectancy chart and to the permanent individual cow record. This permits monitoring of abnormally long cycles and long intervals from freshening to first service.
- Consider using heat detection aids to help increase the number of heats detected. Detection devices and detector animals should supplement routine visual observation.
- In larger herds the use of a testosterone-treated heifer will stimulate more mounting and probably will be cost effective.
- Using prostaglandin to induce estrus in one or two cows will increase the overall estrous behavior in the herd.
- Estrous synchronization programs for the lactating herd, or programmed breeding, will synchronize estrus for several days. Also, more heats will be anticipated and thus more heats are likely to be observed.
- Isolate the cow thought to be in heat with a sexually active cow or heat-detector animal. Heat may not be detected in some cows in a large group situation, but when isolated with an active cow or heifer, a cow possibly in heat may exhibit standing behavior.
- Watch for sexually active groups of cattle. Cows in proestrus or estrus tend to congregate and stay together.
- Adjust the feeding program so that cows calve in proper body condition and weight loss is minimized during lactation.

Table 4. Comparison of estrous cycles of repeat breeding heifers before and during an intensive heat detection program

	Average interval between heats	% short cycles <17 days	% normal 18-24 days	% long cycles >24 days
Source: T. Tanabe and J. Almquist (1960), Penn State Research Bulletin 672. [1] Heifers observed at 8:00 a.m., 4:00 p.m., and midnight.				
Before	48.6	6.3	41.9	51.8

	Average interval between heats	% short cycles <17 days	% normal 18-24 days	% long cycles >24 days
During[1]	20.6	9.9	81.1	9.0

TIMING OF INSEMINATION

Since cattle should be inseminated so that viable sperm are at the fertilization site as the unfertilized egg arrives, it is important to estimate the time of ovulation for each cow that is to be inseminated.

Cow with large ear tag for easy identification.

Ovulation occurs 25 to 32 hours after the onset of standing heat. Standing behavior is the only reliable symptom producers have to determine time of ovulation.

Sperm have to be in the female reproductive tract for approximately six hours before they are capable of fertilizing the egg. This process is termed capacitation. Although live sperm have been found in the female tract up to 48 hours after insemination, sperm viability usually is estimated to be 18 to 24 hours. Improper semen handling or poor insemination technique can dramatically reduce the number of sperm cells available for fertilization and thus can lower the conception rate.

The egg travels very rapidly from the ovulation site to the fertilization site in the oviduct. The fertile life of the egg is shorter than that of the sperm. Ovulated eggs remain fertilizable longer (10–20 hours) than they remain capable of being fertilized and developing into normal embryos (8–10 hours). The likelihood of embryonic death increases as the time beyond this interval increases. Thus viable sperm should be at the site of fertilization awaiting the arrival of the freshly ovulated egg. Breeding either too early or too late allows an aged sperm or an aged egg to interact at the site of fertilization and will result in poor conception. Events and time intervals associated with standing heat and insemination are summarized in Figure 3.

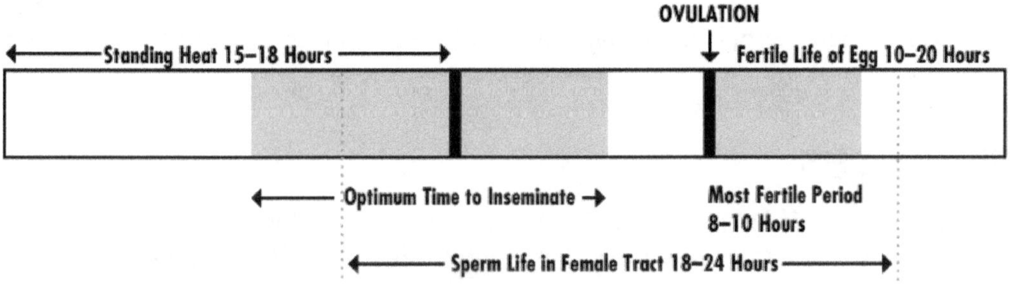

Figure 3. Average time relationships among reproductive events.

Cattle should be inseminated during the last half of standing heat. The a.m.-p.m. rule was developed as a guide. Cows first seen in standing heat in the morning (a.m.) would be inseminated in the afternoon (p.m.) and those observed standing in the evening would be bred the next morning. This system was based on research in which cows were observed frequently (4 to 12 times per day), allowed to interact, and exhibited mounting/standing behavior. Furthermore, insemination was based on standing heat, not secondary signs. Under such conditions, heat detection was very good. But herd managers may not be in a position to accurately predict the latter half of the heat period. Generally it is a challenge just to detect standing behavior. Knowing when to inseminate is another management challenge.

More recent studies conducted by artificial insemination organizations and universities reexamined timing of insemination. In a Virginia study with twice daily heat checks, cows were inseminated either at the end of the heat check period in which they were first observed in heat or at the end of the next heat check period. Using a routine 8:00 a.m. and 8:00 p.m. heat detection system, waiting 12 hours to inseminate resulted in a slight numerical advantage in pregnancy rate over inseminating immediately after heat was first observed (55% vs. 51%). However, this was not a significant difference. Thirty percent of the cows stood to be mounted at the 12-hour heat check after they were first observed in standing heat. These cows had higher pregnancy rates than their herdmates, whether they were inseminated immediately after being first observed in heat or 12 hours later.

Results from a field study in New York showed that near optimal fertility was obtained with a single morning insemination of all cows in heat from the previous evening, including those in heat that morning. This result suggests there may be a benefit to earlier breeding under farm conditions.

Applying this information to a herd situation suggests the following guidelines:

- Best fertility is obtained when cattle are inseminated during the last half of standing heat.
- If a management schedule permits routine heat checks and if it can determine when a heat began and thus predict the latter portion of the heat, the a.m.-p.m. system should be used.
- If the conception rate is unsatisfactory or heat detection is not routine, cows should be inseminated soon after they are first detected in standing heat. Waiting 10 or 12 hours probably results in most of the cows being bred too late.
- Remember, factors other than timing of insemination can affect the conception rate.

REFERENCES

1. Bean, B.H., B.W. Pickett, and R.C. Martig. 1963. Influence of freezing methods, extenders and storage temperatures on motility and pH of frozen bovine semen. J. Dairy Sci. 46:145-149.
2. Berndtson, W.E., B.W. Pickett, and C.D. Rugg. 1976. Procedures for field handling of bovine semen in plastic straws. In: Proc. Nat'l. Assn. Anim. Breeders 6th Tech. Conf. AI and Reprod., Columbia, MO, pp. 51-60.
3. Brackett, B.G., Y.K. Oh, J.F. Evans, and W.J. Donawick. 1980. Fertilization and development of cow ova. Biol. Reprod. 23:189-205.
4. Britt, J.H., R.G. Scott, J.D. Armstrong, and M.D. Whitacre. 1986. Determinants of estrous behavior in lactating Holstein cows. J. Dairy. Sci. 69:2195-2202.
5. Brown, D.W., P.L. Senger, and W.C. Becker. 1991. Effect of group thawing on post-thaw viability of bovine spermatozoa packaged in 0.5 milliliter French straws. J. Anim. Sci. 69:2303-2309.
6. Dalton, J.C., S. Nadir, J.H. Bame, M. Noftsinger, R.L. Nebel, and R.G. Saacke. 2001. Effect of time of insemination on number of accessory sperm, fertilization rate, and embryo quality in nonlactating dairy cattle. J. Dairy Sci. 84:2413-2418.
7. Dalton, J.C., A. Ahmadzadeh, B. Shafii, W.J. Price, and J.M. DeJarnette. 2004. Effect of thawing multiple 0.5-ml semen straws and sequential insemination number on conception rates in dairy cattle. J. Dairy Sci. 87:972-975.

8. DeJarnette, J.M., D.A. Barnes, and C.E. Marshall. 1999. Effects of pre- and post- thaw thermal insults on viability characteristics of cryopreserved bovine semen. Theriogenology. 53:1225-1238.

9. Diskin, M.G., J.R. Pursley, D.A. Kenny, J.F. Mee, and J.M. Sreenan. 2004. The effect of deep intrauterine placement of semen on conception rate in dairy cows. J. Dairy Sci. 87:(Suppl. 1):257 (Abstr.).

10. Dransfield, M.B.G., R.L. Nebel, R.E. Pearson, and L.D. Warnick. 1998. Timing of insemination for dairy cows identified in estrus by a radiotelemetric estrus detection system. J. Dairy Sci. 81:1874-1882.

11. Etgen, W. M., J.M. Ludwick, H.E. Rickard, E.A. Hess, and F. Ely. 1957. Use of mechanical refrigeration in preservation of bull semen. J. Dairy Sci. 40:774-778.

12. Foote, R.H. 1978. Time of artificial insemination and fertility in dairy cattle. J. Dairy Sci. 62:355-358.

13. Hammond, J. 1927. The physiology of reproduction in the cow. University Press, Cambridge, England, pp. 9-49.

14. Hawk, H.W., and T.Y. Tanabe. 1986. Effect of unilateral cornual insemination upon fertilization rate in superovulating and single-ovulating cattle. J. Anim. Sci. 63:551-560.

15. Hawk, H.W. 1987. Transport and fate of spermatozoa after insemination of cattle. J. Dairy Sci. 70:1487-1503.

16. Hopkins, S.M. 1989. Reproductive patterns of cattle. In: Veterinary Endocrinology and Reproduction, Ed. L.E. McDonald, Lea and Fibiger, Philadelphia, PA, pp. 399-415.

17. Hunter, R.H.F., and I. Wilmut. 1983. The rate of functional sperm transport into the oviducts of mated cows. Anim. Reprod. Sci. 5:167-173.

18. Leonard, F.C., J. O'Connell, and K. O'Farrell. 1994. Effect of different housing conditions on behaviour and foot lesions in Friesian heifers. Vet. Rec. 134:490-494.

19. Lopez, H., L.D. Satter, and M.C. Wiltbank. 2004. Relationship between level of milk production and estrous behavior of lactating dairy cows. Anim. Reprod. Sci. 81:209-223.

20. López-Gatius, F. 1996. Side of gestation in dairy heifers affects subsequent sperm transport and pregnancy rates after deep insemination into one uterine horn. Theriogenology. 45:417-425.

21. Maatje, K., S.H. Loeffler, and B. Engel. 1997. Predicting optimal time of insemination in cows that show visual signs of estrus by estimating onset of estrus with pedometers. J. Dairy Sci. 80:1098-1105.

22. Macpherson, J.W. 1968. Semen placement effects on fertility in bovines. J. Dairy Sci. 51:807-808.

23. McKenna, T., R.W. Lenz, S.E. Fenton, and R.L. Ax. 1990. Nonreturn rates of dairy cattle following uterine body or cornual insemination. J. Dairy Sci. 73:1779-1783.

24. Nebel, R.L., W.D. Whittier, B.G. Cassell, and J.H. Britt. 1987. Comparison of on-farm and laboratory milk progesterone assays for identifying errors in detection of estrus and diagnosis of pregnancy. J. Dairy Sci. 70:1471-1476.
25. Nebel, R.L., W.L. Walker, M.L. McGilliard, C.H. Allen, and G.S. Heckman. 1994. Timing of insemination of dairy cows: fixed time once daily versus morning and afternoon. J. Dairy Sci. 77:3185-3191.
26. Nebel, R.L. and S.M. Jobst, 1998. Evaluation of systematic breeding programs for lactating dairy cows: A review. J. Dairy Sci. 81:1169-1174.
27. Nebel, R.L. 2003. Components of a successful heat detection program. Adv. Dairy Technol. 15:191-203.
28. Peralta, O.A., R.E. Pearson, and R.L. Nebel. 2005. Comparison of three estrus detection systems during summer in a large commercial dairy herd. Anim. Reprod. Sci. 87:59-72.
29. Peters, J.L., P.L. Senger, J.L. Rosenberger, and M.L. O'Connor. 1984. Radiographic evaluation of bovine artificial inseminating technique among professional and herdsman-inseminators using 0.5- and 0.25-mL French straws. J. Anim. Sci. 59:1671-1683.
30. Pursley, J.R. 2004. Deep uterine horn AI improves fertility of lactating dairy cows. J. Dairy Sci. 87:(Suppl. 1):372(Abstr.).
31. Pursley, J.R. 2007. Strategies and rationale for resynchronization of ovulation in lactating dairy cows. In: Proc. Dairy Cattle Reprod. Council Annual Meeting, Denver, CO, pp. 37-43.
32. Reimers, T.J., R.D. Smith, and S.K. Newman. 1985. Management factors affecting reproductive performance of dairy cows in the northeast United States. J. Dairy Sci. 68:963-977.
33. Saacke, R.G., J.A. Lineweaver, and E.P. Aalseth. 1978. Procedures for handling frozen semen. In: Proc. Nat'l. Assn. Anim. Breeders 12th Tech. Conf. AI and Reprod., Columbia, MO, pp. 46-61.
34. Saacke, R.G., J.C. Dalton, S. Nadir, R.L. Nebel, and J.H. Bame. 2000. Relationship of seminal traits and insemination time to fertilization rate and embryo quality. Anim. Reprod. Sci. 60-61:663-677.
35. Sangsritavong, S., D.K. Combs, R. Sartori, L.E. Armentano, and M.C. Wiltbank. 2002. High feed intake increases liver blood flow and metabolism of progesterone and estradiol-17β in dairy cattle. J. Dairy Sci. 85:2831-2842.
36. Santos, J.E.P. 2007. Optimization tips and alternatives for timed insemination at first service. In: Proc. Dairy Cattle Reprod. Council Annual Meeting, Denver, CO, pp. 23-35.
37. Senger, P.L., W.C. Becker, S.T. Davidge, J.K. Hillers, and J.J. Reeves. 1988. Influence of cornual insemination on conception in dairy cattle. J. Anim. Sci. 66:3010-3016.
38. Senger, P.L. 1999. The luteal phase of the estrous cycle. In: Pathways to Pregnancy and Parturition, Current Conceptions, Inc., Pullman, WA, pp. 149-166.

39. Stevenson, J.S. 2001. Reproductive management of dairy cows in high milk-producing herds. J. Dairy Sci. 84(E. Suppl.):E128-E143.

40. Trimberger, G.W. and H.P. Davis. 1943. Conception rate in dairy cattle from artificial insemination at various stages of estrus. Neb. Agric. Exp. Sta. Res. Bull. 129:1-14.

41. Trimberger, G.W. 1948. Breeding efficiency in dairy cattle from artificial insemination at various intervals before and after ovulation. Neb. Agric. Exp. Sta. Res. Bull. 153:1-26.

42. Walker, W.L., R.L. Nebel, and M.L. McGilliard. 1996. Time of ovulation relative to mounting activity in dairy cattle. J. Dairy Sci. 79:1555-1561.

43. Williams, B.L., F.C. Gwazdauskas, W.D. Whittier, R.E. Pearson, and R.L. Nebel. 1988. Impact of site of inseminate deposition and environmental factors that influence reproduction of dairy cattle. J. Dairy Sci. 71:2278-2283.

44. Wilmut, I. and R.H.F. Hunter. 1984. Sperm transport into the oviducts of heifers mated early in oestrus. Reprod. Nutr. Develop. 24:461-465.

45. Wiltbank, M.C. 2003. Novel nutritional effects on reproduction. In: Proc. 5th Ann. Intermountain Nutrition Conference, Salt Lake City, UT, pp. 127-140.

46. Xu, Z. Z., D.J. McKnight, R. Vishwanath, C.J. Pitt, and L.J. Burton. 1998. Estrus detection using radiotelemetry or visual observation and tail painting for dairy cows on pasture. J. Dairy Sci. 81:2890-2896.

CHAPTER 7

A.I. Technique in Cattle

Chapter Contents

INTRODUCTION

The recto-vaginal technique is the most commonly used method to artificially inseminate cattle. The basic skills required to perform this technique can be obtained with about three days practice under professional instruction and supervision. Additional proficiency and confidence will be achieved with further work on your own.Several things to keep in mind in choosing a location for inseminating cattle are:

- Safety of both the animal and the inseminator.
- Ease of use.
- Shelter from adverse weather

Regardless of whether you are left or right handed, it is recommended that you use your left hand in the rectum to manipulate the reproductive tract and the right hand to manipulate the insemination gun. This is because the rumen or stomach of the cow lies on the left side of the abdominal cavity, displacing the reproductive tract slightly to the right. Thus, you will find it much easier to locate and manipulate the tract with your left as opposed to right hand. A gentle pat on the rump or a soft-spoken word as you approach for insemination, will help to avoid startling or surprising the animal. Raise the tail with your right hand and gently massage the rectum with the lubricated glove on your left hand. Place the tail on the back side of your left forearm so it will not interfere with the insemination process. Cup your fingers together in a pointed fashion and insert your hand in the rectum, up to the wrist. Gently wipe the vulva with a paper towel to remove excess manure and debris. Be careful not to apply excessive pressure, which may smear or push manure into the vulva and vagina. With your left hand make a fist and press down directly on top of the vulva. This will spread the vulva lips allowing clear access to insert the gun tip several inches into the vagina before contacting the vaginal walls. Insert the gun at a 30° upward angle to avoid entering the urethral opening and bladder located on the floor of the vagina. With the gun about 6 to 8 inches inside the vagina, raise the rear of the gun to a somewhat level position and slide it forward until it contacts the external portion of the cervix. You will note a distinct gristly sensation on the gun when it contacts the end of the cervix. The cervix, consists of dense connective tissue and muscle and is your primary landmark for inseminating cattle. It has often been described as having the size and consistency of a turkey neck. The size will vary, with post partum interval and age of the animal. The cervix usually has three or four annular rings or folds. The opening into the cervix protrudes back into the vagina. This forms a 360° blind-ended pocket completely around the cervical opening. This pocket is referred to as the fornix. In most cows, the cervix will be located on the floor of the pelvic cavity near the anterior end of the pelvic bone. In older cows with large reproductive tracts, the cervix may rest slightly over the pelvic bone and down into the abdominal cavity.

To become a successful inseminator it is very important that you always know where the tip of the insemination gun is located. The walls of the vagina consist of thin-layered muscle and loose connective tissue. The insemination gun can be easily felt with your palpating hand. As you insert the breeding gun into the vagina, keep your gloved hand even with the gun tip. Manure in the rectum can often interfere with your ability to palpate the cervix and gun tip. However, it is seldom necessary to remove all the manure from the bowel. Instead, keep your open hand flat against the floor of the rectum, allowing the manure to pass over the top of your hand and arm. While handling the cervix you may notice rectal constriction rings attempting to force your arm from the cow. To relax these rings, place two fingers through the center of a ring and massage back and forth. The constriction ring will eventually relax, pass over your hand and arm, and you can continue the palpation process. Because the reproductive tract is freely movable, cows that have strong rectal and abdominal contractions in response to being palpated may actually push their reproductive tract back into the pelvic cavity. This will cause many folds to form in the vagina. In such cases, the insemination gun will often get caught in these folds and little or no progress will be made until they are removed. If you can locate the cervix, grasp it and push it forward. This will straighten the vagina and the gun should pass freely up to the cervix. If you cannot locate the cervix, encircle the gun tip with your thumb and forefingers. With a straightening motion of your wrist, gently "milk" the folds out of the vagina a little at a time. Slide the gun forward and repeat the process until the cervix is reached.

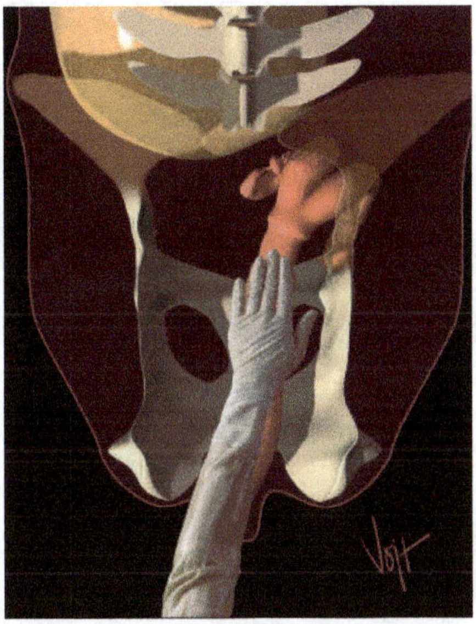

Because the rumen displaces the reproductive tract to the right, it is much easier to locate and manipulate the tract with your left hand.

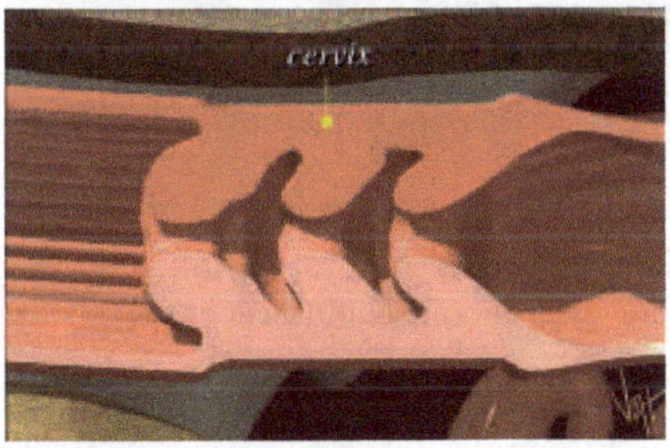

The opening into the cervix protrudes back into the vagina

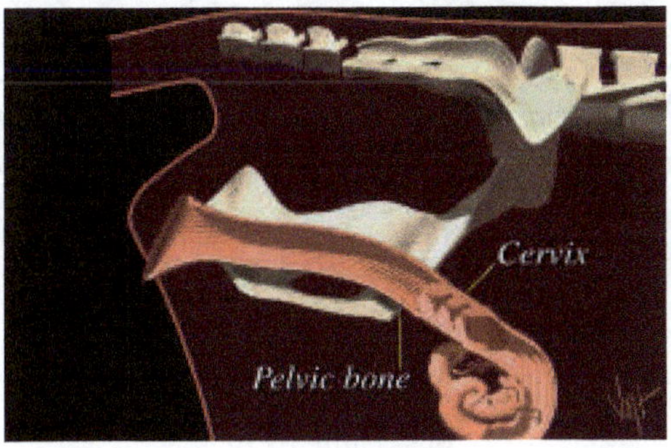

The cervix is located on the floor of the pelvic cavity near the anterior end of the pelvic bone.

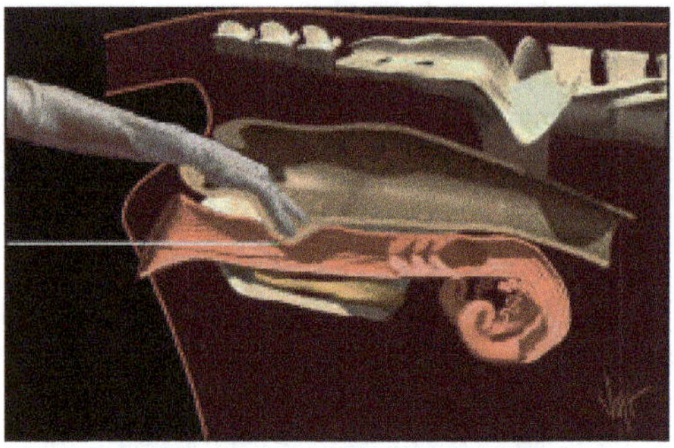

As you insert the breeding gun into the vagina, keep your gloved hand even with the gun tip.

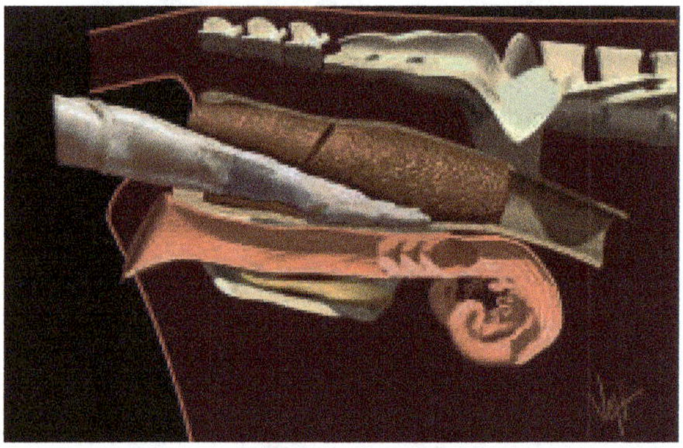

Keep your open hand flat against the floor of the rectum, allowing manure to pass over the top of your hand and arm.

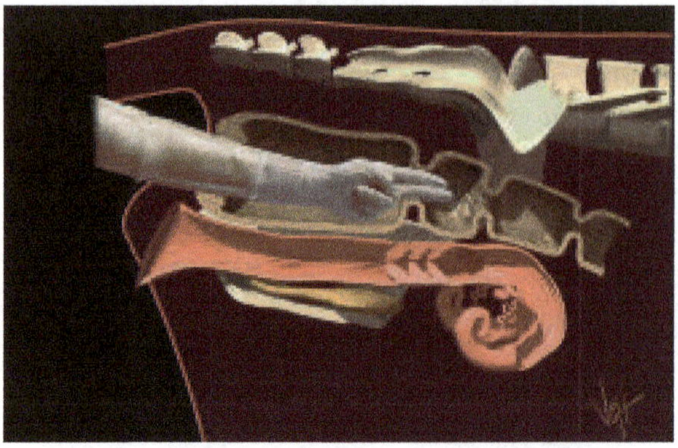

To relax rectal constriction rings, insert two fingers through the center of the ring and massage back and forth.

At this point it is important for you to understand that inseminating a cow is a twostep process. The first step is to get the gun tip to the cervix. To accomplish this you must work the vagina and cervix forward, away from you to straighten the vaginal folds. If you do not feel the gristly sensation of the cervix on the gun, you are still in step one of the process. Once the gun is in contact with the external surface of the cervix you are ready to begin step 2. In step 2, you place the cervix on or over the insemination gun. That's right, the cervix is placed on the gun, the gun is not passed through the cervix. Excessive movement or probing with the insemination gun during the second step is seldom productive and in fact, is very often counterproductive. Ground gained is often lost and we find ourselves back in a vaginal fold. The key to mastering step 2 of the insemination process is to know how to hold and manipulate the cervix and concentrating on doing the work with the hand inside the cow, not the one holding the gun. When the gun first contacts the cervix, you will usually find that the tip is in the fornix directly over top of the opening. Grasp the external opening to the cervix with the thumb on top and forefingers underneath. This closes the fornix at top and bottom. As in step 1 we must still know the location of the gun tip. This is accomplished with the palm and third and fourth fingers of your palpating hand. Use your palm and these two fingers to guide the gun tip to the cervical opening located between your thumb and forefingers.

With gentle probing the opening should be located. You will feel the gun slide forward until it contacts the second cervical ring. Maintain gentle but steady forward pressure on the gun and slide your thumb and forefingers just in front of the gun tip and re-grasp the cervix. Because the cervix is composed of dense connective tissue and muscle, it is difficult to clearly distinguish the gun tip when it is located within this structure. However, you can determine the approximate location by bending the cervix. Using the flexibility of your wrist, twist and bend the cervix until you feel the second ring slide over the gun tip. Repeat the process until all the rings have been passed over the gun tip. In some cases, it may be necessary to bend the cervix at a 90° angle to clear the cervical folds. Remember, you are placing the cervix over the gun, not the gun through the cervix.

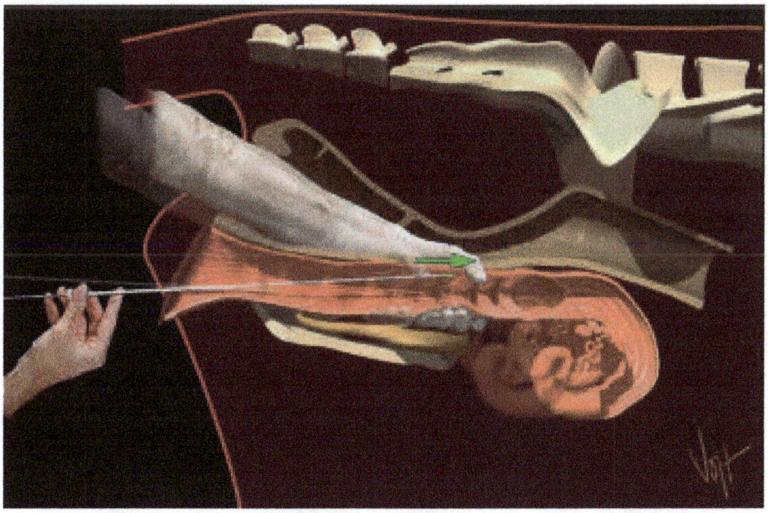

Grasp the cervix and push it forward to straighten vaginal folds.

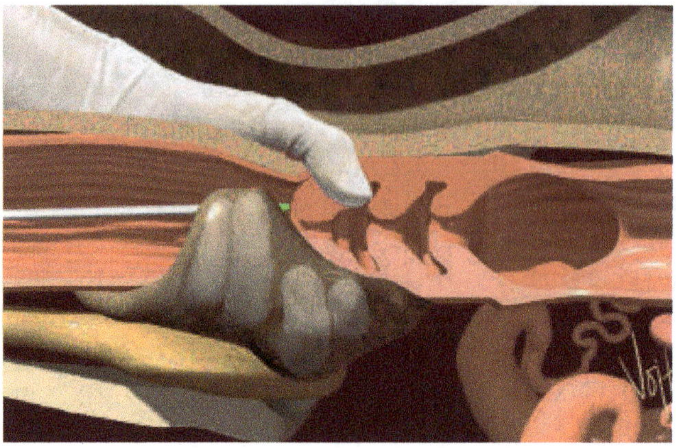

Grasp the external opening to the cervix with the thumb on top and the forefingers underneath to close the fornix and guide the gun tip into the cervix.

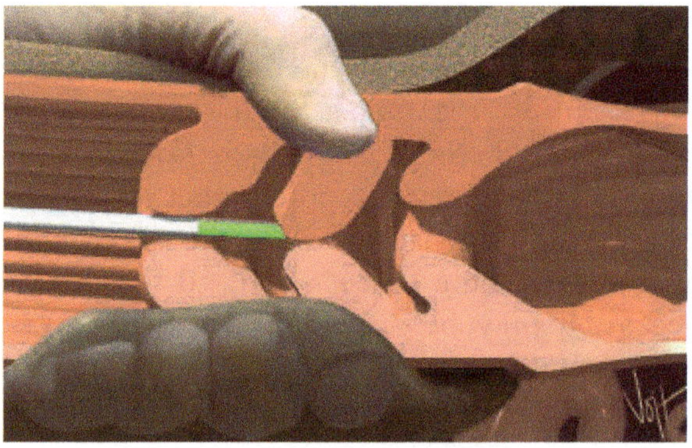

Using the flexibility of your wrist, twist and bend the cervix until you feel the second ring slide over the gun tip.

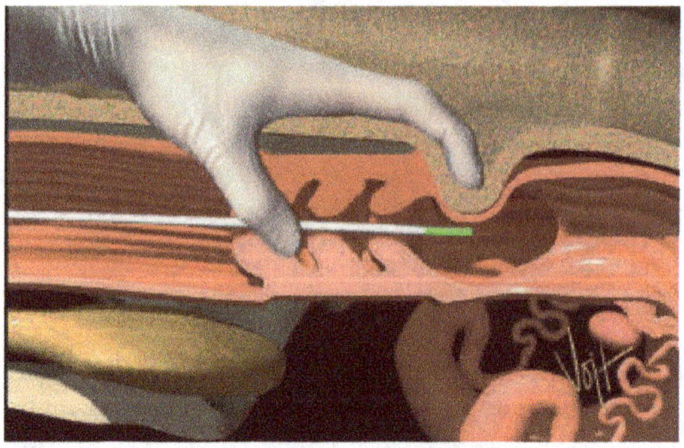

Use your index finger to check gun placement (1/4 inch past the end of the cervix) before depositing semen.

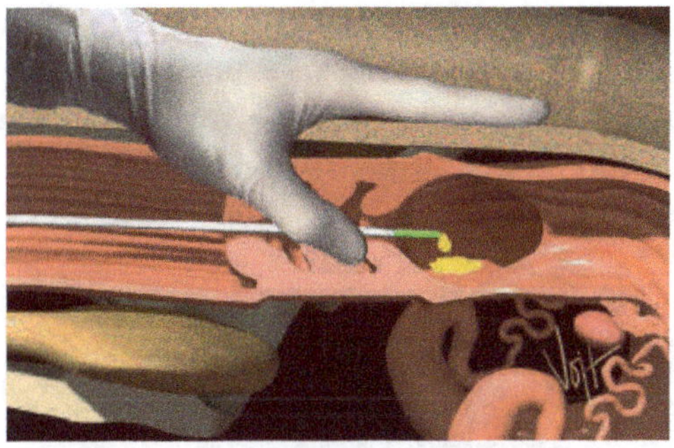

Push the plunger slowly so that drops of semen fall directly into the uterine body

A slight jiggle or "give-and-go" of the gun may sometimes be necessary to help pass by a fold, but for the most part gentle forward pressure is all that is necessary and gun movement should be minimal. When all rings of the cervix have been cleared, the gun should slide forward freely with little resistance. Since the uterine wall is very thin, you will once again be able to clearly feel the insemination gun. You are now ready to check your placement and deposit the semen. Rotate your gloved hand until it lies on top of the cervix. With your index finger, locate the far end of the cervix. Pull back on the gun until you feel the tip directly underneath your finger near the internal opening of the cervix. Raise your finger and slowly deposit the semen. Push the plunger slowly so that drops of semen fall directly into the uterine body. With proper A.I. technique and gun placement, semen will be deposited in the uterine body. Uterine contractions will then transport spermatozoa forward to the horns and oviducts with a good distribution of both sides.

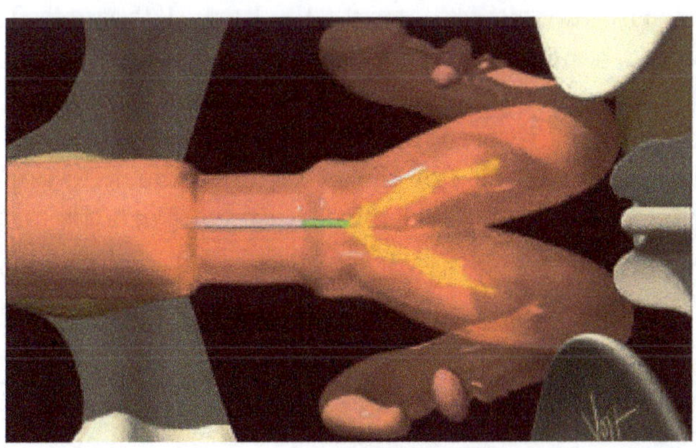

With proper A.I. technique and gun placement, semen will be deposited in the uterine body and contractions will transport spermatozoa forward to the horns and oviducts.

When the insemination gun is more than 1" through the cervix, all the semen will be deposited in only one horn. This creates a situation of uneven semen distribution. Should the animal ovulate from the opposite horn, conception rates may be compromised. Be sure to raise your finger after checking gun placement. Not doing so may obstruct one horn, again creating a situation of uneven semen distribution. When checking gun tip placement, be careful not to apply excessive pressure. The delicate uterine lining is easily damaged, predisposing the cow to uterine infections and reduced fertility. Make sure you push in with the plunger and do not pull back on the gun. Pulling back may result in much of the semen dose being deposited in the cervix and vagina instead of the uterine body.

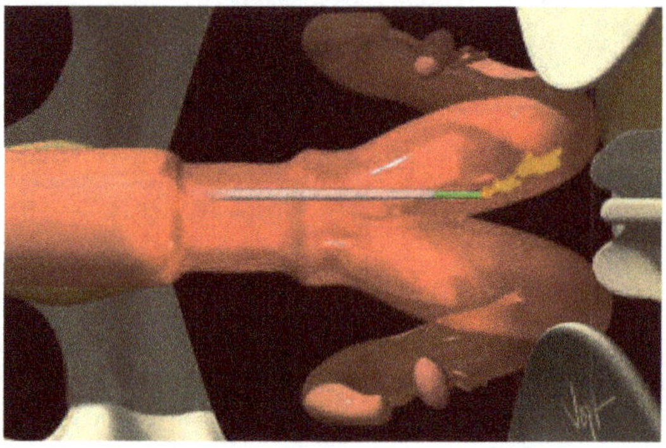

If the gun is more than 1 inch through the cervix, all the semen will be deposited into only one horn.

Although the recommended site of semen deposition is in the uterine body, research suggests that when exact gun tip placement is in doubt, depositing semen slightly into one uterine horn is less likely to compromise fertility than cervical deposition. However, If the cervical mucous of a cow which has been previously inseminated feels thick and sticky on the gun, she may be pregnant. In this case deposit the semen about halfway through the cervix. After properly depositing semen, slowly pull the gun from the reproductive tract. Remove the gloved hand from the rectum and shake off the excess manure. Check the gun tip for signs of blood, infection or semen leakage inside the sheath. Make notes for your veterinarian or future reference where appropriate. Remove the sheath from the gun and hold it in the gloved hand. For the final time, check to confirm which bull you have used. Remove the glove starting at the top of your arm by turning it inside out as you remove it. Remove air from the glove and tie a knot at the open end to trap manure, the sheath and dirt inside. Dispose of the used glove in a proper receptacle. Wipe the gun clean and dry and return it to the proper storage location. Some of the most important aspects to remember when inseminating a cow to attain maximum breeding efficiency are:

- Be gentle. Don't use too much force.
- Insemination is a two-step process. Get the gun to the cervix, and then
- place the cervix over the gun.
- Deposit the semen just through the cervix into the uterine body.
- Take your time.

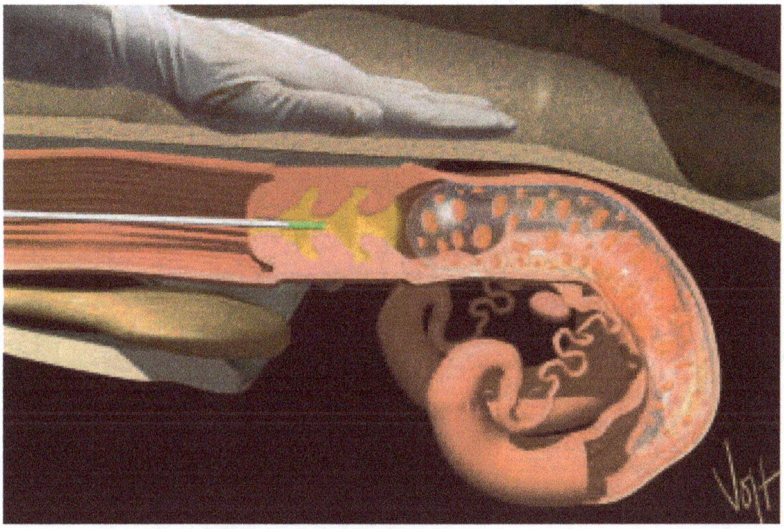

If you encounter cervical mucous which feels thick and sticky on the gun in a cow that has been previously insemi- nated, she may be pregnant. In this case, deposit the semen halfway through the cervix.

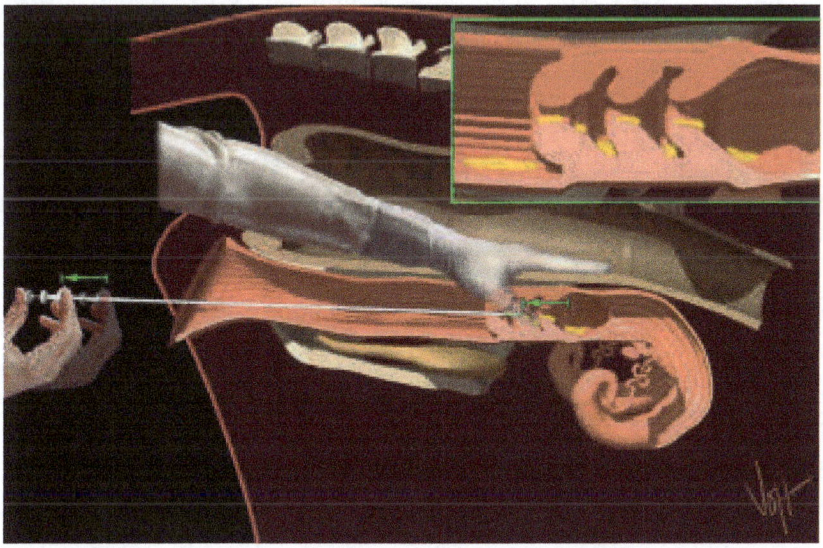

Make sure you push in with the plunger and do not pull back on the gun. Pulling back may result in much of the semen dose being deposited in the cervix and vagina.

Correct insemination procedures will result in better breeding efficiencies. More selection pressure can then be placed on economic traits such as milk and beef production enabling you and your family to realize a higher return on your semen investment dollars. υ

REFERENCES

1. Asare, B.P. 1970. Potentialities for fodder production in the humid zone of Ghana. Proc. Ghana Anim. Sci. Assoc., 3: 42-47.
2. Davis, H.P., Plum, M. & Brost, B. 1954. Studies on the herd management records. II. Relations of gestation length to birth weight of Holstein calves in both sexes at various carvings. J. Dairy Sci., 37: 162-169.
3. Diggins, R.V., Bundy, C.E. & Christenson, V.W. 1984. Dairy production (5th ed.). Englewood Cliffs, New Jersey, Prentice-Hall, Inc.
4. Gyawu, P. & Agyemang, K. 1977. The performance of a herd of Holstein-Friesians in the humid tropical forest zone of Ashanti, Ghana. Proc. Ghana Anim. Sci. Assoc., 9: 12-19.
5. Hernandez, P.A. 1965. A study of some reproductive characteristics in pure-bred and crossbred cows in Venezuela. Anim. Breeding Abstr., 34: 222.
6. Kabuga, J.D. & Alhassan, W.S. 1981. The reproductive performance of a small herd of Holstein/Friesians in the humid forest zone of Ghana. World Rev. Anim. Prod., 17 (3): 41-48.
7. Kassir, S.A., Juma, K.H. & Al Jaff, P.H. 1969. A further study on dairy characters in Friesian and crossbred cattle in Iraq. Trop. Agric. (Trinidad), 46: 359-363.
8. Mahadevan, P. 1956. Breeding for milk production in tropical cattle. Tech. Bull., No. 17. Farnham Royal, UK, Commonwealth Bureau of Animal Breeding and Genetics.
9. Martinez, N., Galina, S. & Combellas, J. 1982. Reproductive and productive performance of a Holstein herd in Maracay, Venezuela. Trop. Anim. Prod.. 7: 31-39.
10. McDowell, R.E. 1972. Improvement of livestock production in warm climates. San Francisco, W.H. Freeman and Co.
11. Otchere, E.O. 1976. Cattle feeding in Ghana - a review. Proc. Ghana Anim. Sci. Assoc., 8: 2-11.
12. Payne, W.J.A. 1951. Records From the dairy herd at Singatoka for the decade 1941-1950. Anim. Breeding Abstr., 20: 1 006.
13. Rendel, J. 1972. Dairy cattle in hot climates. World Rev. Anim. Prod., 8: 16-24.
14. Ridler, B., Broster, W.H. & Foot, A.S. 1963. The growth rate of heifers in a dairy herd. J. Agric. Sci., 61: 1-8
15. Spalding, R.W., Everett, R.W. & Foote, R.H. 1975. Fertility in New York artificially inseminated Holstein herds in dairy herd improvement. J. Dairy Sci., 58: 718-723.
16. Touchberry, R.W. 1967. A study of the N'Dama cattle at the Mussaia Animal Husbandry Station in Sierra Leone. Univ. Illinois Agric. Exp. Stn. Bull., No. 724.
17. Trail, J.C.M. & Marples, H.J.S. 1968. Production traits for lactations for Friesian cattle of Uganda. Trop. Agric. (Trinidad), 45: 173-178.
18. Wilkins, J.V. 1986. Productive and reproductive performance of cattle in the

tropics. In Nuclear and related techniques in animal production and health. Vienna, International Atomic Energy Agency.

CHAPTER 8

Pregnancy Diagnosis

Chapter Contents

The accurate diagnosis of pregnancy is of crucial importance in the establishment and maintenance of optimal reproductive performance. It is desirable for the farmer to know as soon as possible if a mated cow is not pregnant, so that she can be rebred with the minimum delay. Until fairly recently pregnancy diagnosis consisted of observation for oestrus, especially at about 21 days after service. Given that oestrus detection efficiency and accuracy are low in most herds, reliance on this method results in a large number of nonpregnant cows assumed to be pregnant and some pregnant cows being inseminated inappropriately, which could abort them. Furthermore, a large number of cows sold as barren are found, at slaughter, to have been pregnant. A number of other methods have been developed over the years, and there follows a description of potentially useful methods and suggestions for their optimal use in cattle fertility management.

METHODS OF PREGNANCY DIAGNOSIS

Non-Return to Service

Traditionally, a cow was diagnosed as non-pregnant if she was seen in oestrus (and, as a rule, re-inseminated) approximately 21 days after service. The percentage of cows not seen in oestrus at about this time is known as the nonreturn rate and is an established method of estimating pregnancy rates. It is usually the first method (in terms of time after service) used by the stockperson to discriminate between non-pregnant and pregnant cows. Since the percentage of ovulating cows actually seen in oestrus is often as low as 50%, the non-return rate usually overestimates pregnancy rates and, in the individual cow, can easily lead to a false sense of security. However, an effective procedure for detecting oestrus in non-pregnant cows is still a very important aspect of good reproductive management. It is useful to have a recording system which will draw attention to cows due to return, but, of course, this should not be to the detriment of observation at other times in case of returns at abnormal intervals. If a cow is recorded in oestrus approximately three weeks after service, it is still possible that she could be pregnant. Inseminating her is very likely to cause abortion. Great care should be taken to confirm that she is genuinely in heat and not pregnant. A 'cow-side' milk progesterone test (see below) could be used and insemination avoided unless the progesterone level is very low. If insemination is carried out and there is any possibility of an existing pregnancy, then the AI pipette should not be passed all the way through the cervix. Oestrous behaviour and oestrus detection.

Progesterone Measurement

The detection of pregnancy by measurement of hormones, particularly if they occur in milk, has an advantage over rectal palpation in that interference with the cow is minimal and there is a negligible risk to the pregnancy. The measurement of progesterone to check for pregnancy also offers the possibility of diagnosis at day 21 or even earlier. Progesterone is secreted by the corpus luteum throughout pregnancy. As was discussed in Chapter peak values are usually reached by about day 10 after ovulation and are subsequently maintained, whereas in the nonpregnant animal levels fall at about day 17 of the cycle. Consequently, in the pregnant cow, blood and milk progesterone concentrations remain high between days 21 and 24 after ovulation, when they would be basal in the nonpregnant animal. Therefore a sample taken at this time may be used to discriminate between the pregnant and non-pregnant cow. The development of rapid highly sensitive ELISA (enzyme-linked immunosorbent assay) procedures has allowed the commercial exploitation of this test. The introduction of monoclonal, rather than polyclonal, antibodies (Groves et al., 1990) lowered the limits of detection and increased the discrimination of the assays. The simplest procedure is to take a milk sample from the cow at about 20 days after insemination. If the progesterone level is low, the cow is definitely not pregnant, provided that the right sample has been taken from the correct cow and the test has worked properly. If the progesterone level is high, the cow could be pregnant.

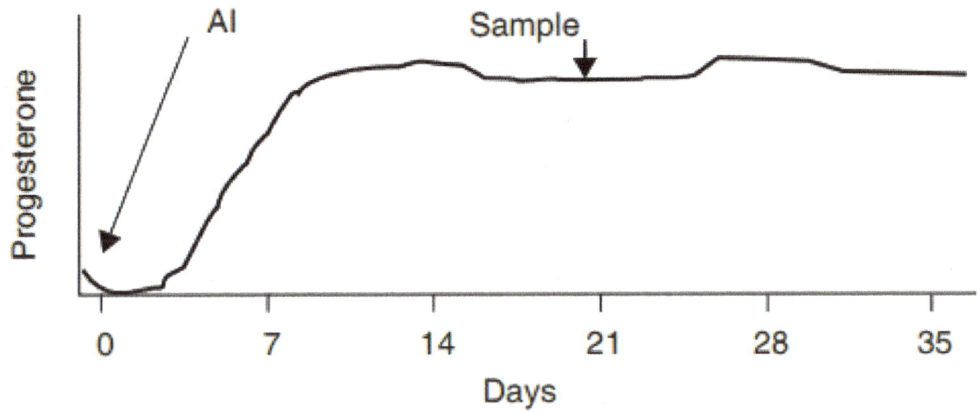

Schematic representation of correct positive milk progesterone pregnancy diagnosis.

The exact day to sample could range from 19 (when a low progesterone value would signal a careful watch for oestrus, enabling a prompt return service) to 24 (giving the cow a chance to show heat on return, so that there would be no need to sample). It is important to obtain a bulk milk sample since the progesterone concentration is dependent on the level of milk fat, which is in low concentration in foremilk and high concentration in strippings. This is of obvious importance if the method is used for beef suckler cows since the time of sampling in relation to the calf suckling can affect the result. The test could also be carried out after blood sampling in non-lactating animals such as maiden heifers. This may stress the animal and would involve veterinary intervention and its associated costs. The concentration of progesterone reflects the function of the corpus luteum, but not necessarily the presence of an embryo or fetus. A high progesterone level suggests that the cow is pregnant, but a single test like this is only about 80% accurate for positive pregnancy diagnoses. There are a number of reasons for false positive tests:

- The cow may not be pregnant but the life of the corpus luteum may still be extended for some reason, such as a uterine infection.
- The cow could have been inseminated at the incorrect stage of the cycle, i.e., during the luteal phase. At the time of the test, she is likely to be in the luteal phase of the subsequent cycle, and the progesterone level would be high. A sample taken at the time of AI would help to confirm that the timing was right and improve the accuracy of the pregnancy diagnosis sample.
- The diagnosis may be accurate, but the embryo or fetus is lost afterwards. This is now quite common – see Chapter. A further sample 42 days after AI may fail to show that this has occurred since ovarian cycles may have been re-established, making it likely that a high progesterone level would be measured on the day of the subsequent sample.

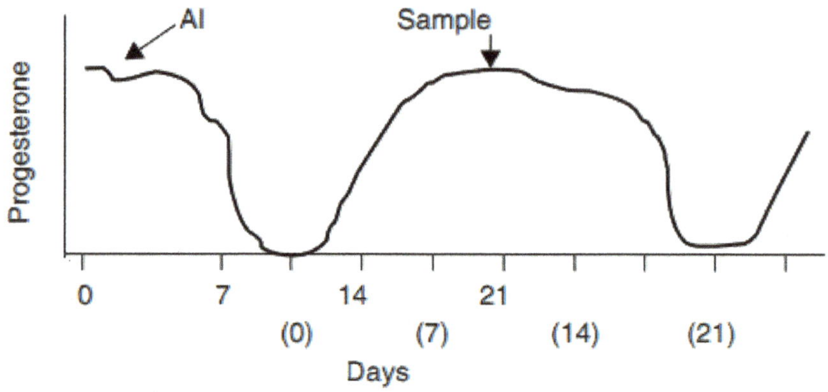

Schematic representation of false positive milk progesterone pregnancy diagnosis in a cow inseminated at the wrong time.

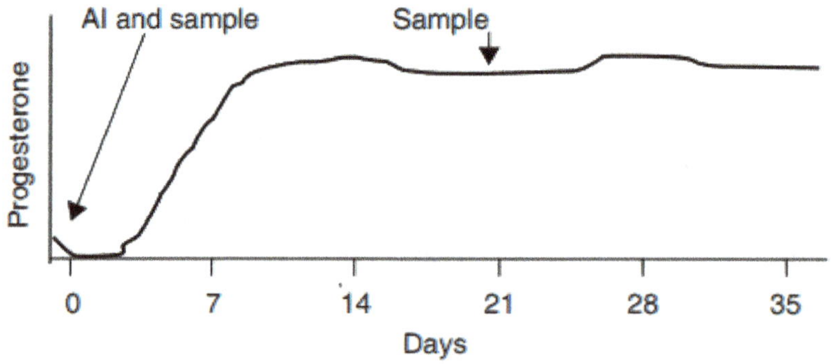

Schematic representation of correct positive milk progesterone pregnancy diagnosis.

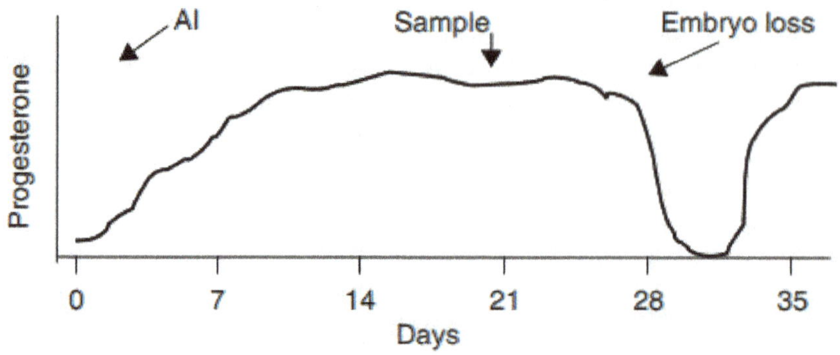

Schematic representation of correct positive milk progesterone pregnancy diagnosis, but in a cow that subsequently lost her conceptus.

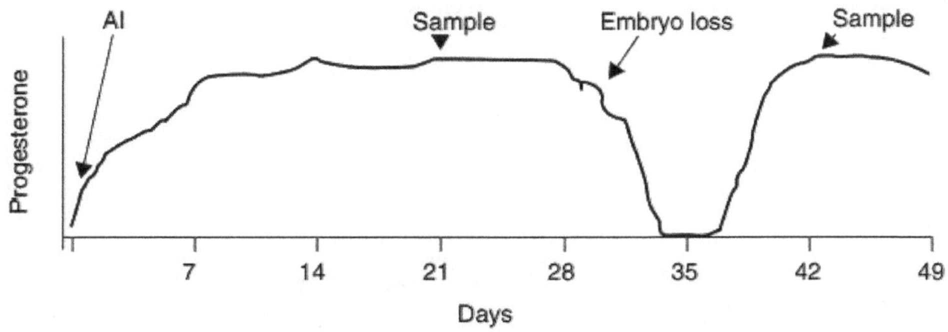

Schematic representation of milk progesterone pregnancy diagnosis in which a second sample failed to reveal an embryo loss.

Other reproductive problems, such as luteinized cystic follicles, could lead to high progesterone levels on the day of the pregnancy diagnosis test.

In summary, approximately 80–85% of positive diagnoses between days 21 and 24 appear to be accurate, i.e., they result in a calf being born at the appropriate time. Therefore 15–20% of cows with high progesterone levels at days 21–24 do not calve subsequently, i.e., they are false positives. In contrast, the finding of a low progesterone level between days 21 and 24 indicating nonpregnancy is likely to be accurate in almost 100% of cases. The cow may have failed to conceive and returned normally or may have inactive ovaries, or cystic follicles, so that no progesterone is being produced. Other types of cystic ovarian disease cause erratic progesterone production, so that levels could be low on the day of the test. These problems are described in Chapter.

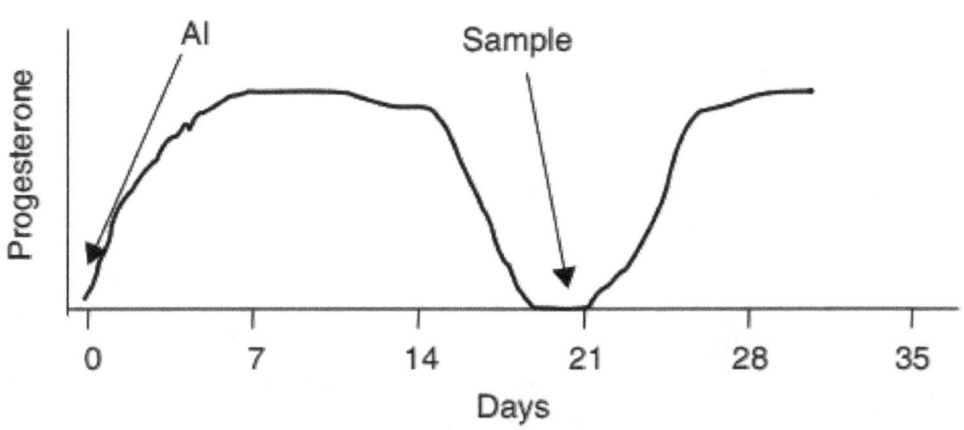

Schematic representation of correct negative milk progesterone pregnancy diagnosis.

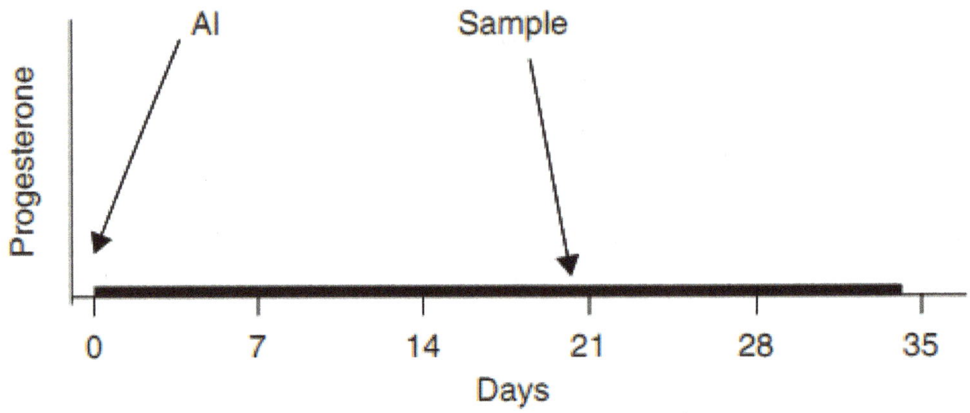

Schematic representation of correct negative milk progesterone pregnancy diagnosis in a cow with inactive ovaries or cystic follicles.

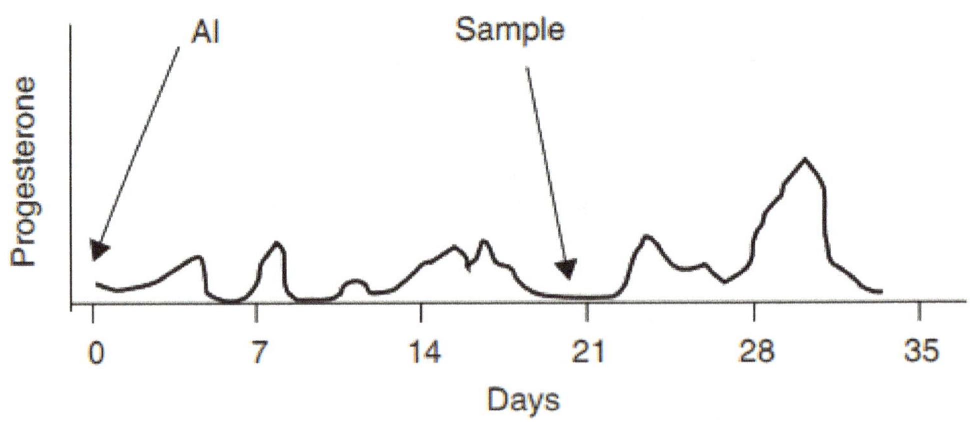

Schematic representation of correct negative milk progesterone pregnancy diagnosis in a cow with luteinized cystic follicles.

In any event, a low progesterone level in a correctly sampled and tested milk sample means that the cow will certainly not be pregnant. Thus, milk progesterone measurement may be considered as a highly accurate test for nonpregnancy and can be very useful in allowing early rebreeding, providing cycles are normal. In fact, as was shown in Chapter, it is possible to use a series of milk samples taken from about day 18 after insemination not only to detect pregnancy failure, but also to determine the time to re-inseminate, even if oestrus is not observed. Milk progesterone pregnancy testing was first applied in the UK by the Milk Marketing Board in the mid 1970s. A single bulk milk sample was taken between days 21 and 24 after insemination and despatched to the laboratory by post for assay. The associated delay in receiving results tended to cancel out the advantages of testing early after insemination. The development of a test that could be carried out on farm (commonly known as the 'cow-side test') enabled results to be obtained rapidly and thus be of more potential use in reproductive management. A number of 'cow-side' progesterone test kits are now available, offering the possibility of results within hours, or even minutes, of sampling at very cost effective prices. However, it is currently estimated that less than 5% of dairy cows in Great Britain are tested using the milk progesterone method. The development of an automated assay to measure individual cow progesterone levels at milking is eagerly awaited by many farmers with large herds and computerized milking parlours. The development of such assays is discussed in Chapter. As was seen in Chapter, cows with lower progesterone levels as soon as day 5 after insemination are likely to have lower pregnancy rates. Marshall et al. (2003) found that milk progesterone levels on day 6 after insemination were almost twice as high in cows subsequently diagnosed pregnant as in those diagnosed non-pregnant. Even so, individual variation within both pregnant and non-pregnant cows means that there is little chance of using this measurement as a very early diagnostic test. They also measured an increase in milk progesterone levels from day 5–6 of 3 ng/ml in subsequently pregnant cows compared to 1 ng/ml in non-pregnant cows and suggest that, if the results were repeated with larger numbers of cows, this could be a possible means of making a tentative, but potentially useful, very early diagnosis of pregnancy.

Pregnancy-Specific Proteins

Early pregnancy factor (EPF) is a pregnancy-dependent protein complex that has been detected in the serum of several species (Morton et al., 1983). It is detected using an immunological technique, the rosette inhibition test. It is claimed that this substance appears in serum shortly after conception and disappears very rapidly following embryonic death. Therefore the detection of EPF would offer a very valuable method of pregnancy diagnosis. Whilst a degree of success has been achieved using this technique in some laboratories (e.g., Yoshioka et al., 1995), results in the cow have been insufficiently reproducible to offer much immediate hope of an improved method of pregnancy diagnosis. Two further proteins, bovine pregnancy-specific protein B (bPSPB) and bovine pregnancy associated glycoprotein (bPAG – an antigen synthesized in the superficial layers of the ruminant trophoblast), can be measured during early pregnancy in the cow with much more reliability. Szenci et al. (1998b) found that the use of either of these proteins compared favourably with ultrasound pregnancy at days 26–58 after AI, although the accuracy of detecting non-pregnant animals by both protein tests was limited by the relatively long half-life of these proteins after calving and by early embryonic mortality. They reported significantly fewer false positive diagnoses by the bPSPB test than by the bPAG test. Patel et al. (1997) reported that peripheral bPAG concentrations in Holstein cows after non-surgical embryo transfer were correlated to the stage of gestation and fetal number, and that the profile of the peripheral plasma concentrations provided a useful indication of feto-placental status. Humblot (2001) has reviewed the use of pregnancy-specific proteins to monitor pregnancy and study late embryo mortality in ruminants. bPAG has been produced commercially but is currently not available in the UK. The main drawbacks to its routine use in pregnancy diagnosis are that it is only detectable in blood and not milk, plus the carryover from the previous pregnancy and the failure of levels to fall promptly after conceptus loss (e.g., Szenci et al., 1998a).

Real-Time Ultrasound Scanning

The availability of real-time b-mode ultrasound scanning has been a major advance in pregnancy diagnosis and reproductive monitoring in a number of species including man, sheep, dogs, horses and cattle. It has the advantage that it is non-invasive apart from the requirement in the large farm species for insertion of a rectal transducer, and thus it is a very safe procedure. Pregnancy can be detected as early as 17 days using this method. A very comprehensive account of the use of ultrasound to examine the bovine reproductive tract is given by Boyd (1991), but a brief summary of the detectable changes in early pregnancy is worthwhile here. The conceptus begins to occupy the contralateral horn by about day 17 after service and it can be observed as a non-echogenic (dark) area. By day 19 the amniotic sac has expanded considerably and therefore the lumen of the uterus can readily be observed. By day 22 it is possible to see the beating embryonic heart and by day 30 the conceptus is so obvious that a very rapid and accurate diagnosis can be made. By day 35 the uterine caruncles can be visualized individually. Later, fetal age can be determined approximately by the measurement of crown rump length. A series of ultrasound images of various stages of pregnancy is shown in Fig.

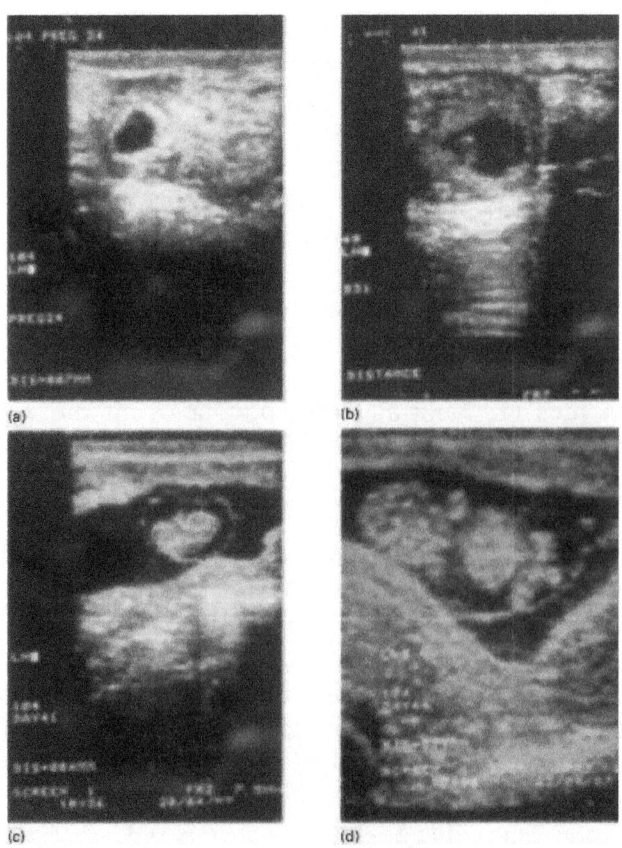

A series of ultrasonographs of the pregnant uterus on days (a) 24, (b) 31, (c) 41 and (d) 44. The uterine lumen can be seen clearly as an anechoic (dark) area and the developing embryo as an echogenic (lighter) structure inside the lumen, increasing in size and definition with age. The placental membranes are clearly visible in chapters (c) and (d) and the limb buds in (d) (day 44). (Reproduced by kind permission of Prof. J.S. Boyd and Drs S.N. Omran and N. Mowa.)

This method of pregnancy diagnosis has a double advantage in that it can be carried out relatively early – in practice from about 28 days postinsemination – and that positive diagnoses are highly accurate. Furthermore, it is sometimes possible to detect problems with the pregnancy and predict losses. For example, an infection sometimes shows as a cloudiness in the amniotic fluid, which would normally give a clear black echo. If the scan is carried out between four and five weeks after insemination, any cow or heifer that failed to conceive is likely to be in the middle of the luteal phase of the second cycle after insemination and, if it is absolutely certain that she is not pregnant, she could be synchronized with prostaglandin for a fixed-time re-insemination three and four days later. We have used the procedure successfully in synchronized groups of maiden heifers at the Scottish Agricultural College (SAC). It has been suggested that early scanning may actually cause loss, but in a large-scale study at SAC we found no evidence of this.

Rectal Palpation of the Fetus

This technique relies on the ability to feel the presence of a fetus swelling in one of the uterine horns by inserting an arm into the rectum of the cow. This can be a hazardous process since trauma can be inflicted on both cow and fetus and should therefore only be carried out by a trained operator. In the nonpregnant and early pregnant animal the uterine horns can be felt to be approximately equal in size and diameter. It is possible to detect a difference in the size of the two horns from about day 40 of pregnancy onwards. This is easier in heifers than in multiparous cows and is to a large extent dependent on the skill of the operator. If, from about day 30, a fold of the uterine wall is picked up and then released, the sensation of two layers of tissue passing through the fingers may be felt. This so-called membrane-slip is due to the presence of the chorioallantois inside the horn. It is advised that this technique is not practised routinely, since it has been known in some cases to damage the fetus and to result in fetal death and abortion. However, following a survey of 7500 cows, it was concluded that this technique did not significantly affect the outcome of the pregnancy (Vaillancourt et al., 1981). The outcome obviously depends on the skill of the individual perfoming the palpation. From 6–8 weeks the disparity in horn size becomes quite marked, the pregnant horn becoming up to six times the diameter of the non-gravid horn at 10 weeks. After this stage the non-gravid horn also begins to enlarge due to the invasion and attachment of the placenta throughout the uterine lumen. By 12 weeks the diameter of the pregnant horn may be 10 cm or more and is easily detectable. As the gravid uterus becomes larger and heavier it begins to sink down below the pelvic brim and by the fifth month it may not be possible to palpate it. Only when the fetus has grown further, to the seven-month stage, may it be

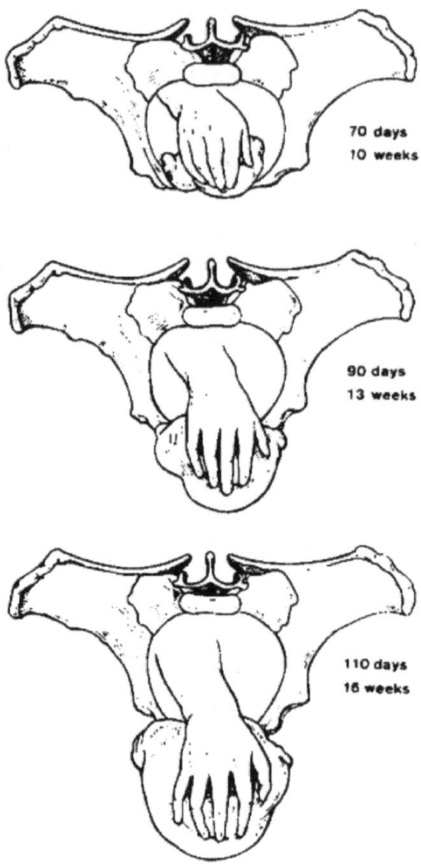

Detection of pregnancy by rectal palpation. (Redrawn from Arthur et al., 1982.)

possible to feel it again. However, in these cases pregnancy may be suspected by the absence of the normal reproductive tract in the pelvic cavity and the palpation of the cervix and anterior vagina as a taut band of tissue passing over the pelvic brim. The presence of cotyledons may or may not be felt from about the fourth month onwards. The sensation is quite characteristic and has been described as like 'corks bobbing on water'. After the seventh month of gestation the fetal head can usually be palpated with ease. Another indication of pregnancy may develop about mid-term. Blood flow through the middle uterine artery on the pregnant side increases dramatically during pregnancy. This usually pulsates like a normal artery, but in advanced pregnancy slight compression of this reveals a characteristic vibrating sensation known as fremitus. Certain pathological conditions of the uterus can be confused with a normal pregnancy by causing uterine enlargement. These include pyometra, large tumours and the presence of a mummified fetus. The detection of pregnancy by rectal palpation is described in more detail by Noakes (1991).

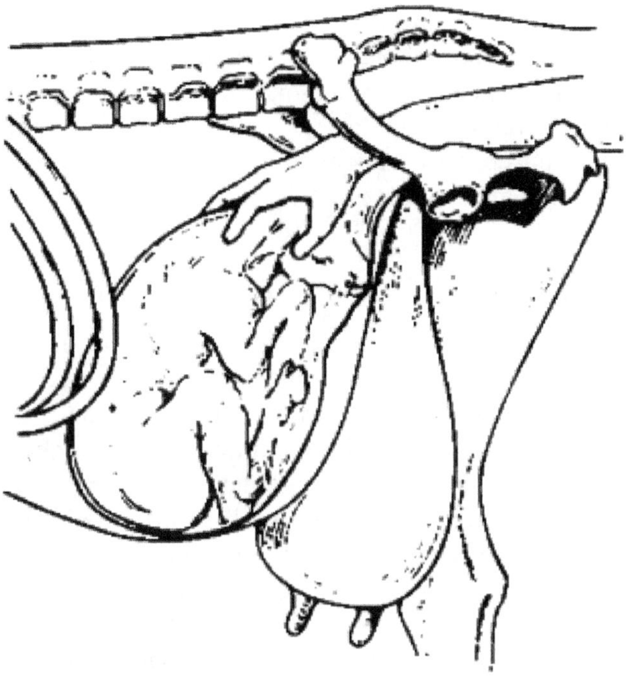

Detection of pregnancy by rectal palpation. Pregnancy approaching term. (Redrawn from Arthur et al., 1982.)

OESTRONE SULPHATE

The identification of compounds produced by the conceptus rather than by the dam would have clear advantages in the diagnosis of pregnancy in farm animals. Oestrogens are produced by the bovine conceptus and oestrone sulphate concentrations in maternal plasma rise from about day 70 of pregnancy. The oestrone sulphate content of milk reflects that of plasma and it has been found that a positive oestrone sulphate test at 15 weeks of pregnancy gives a 100% accurate diagnosis of pregnancy. In most cases, detection of non-pregnancy at this late stage is of limited value to reproductive management, but it can be used as a confirmatory test following an earlier positive diagnosis, such as a milk progesterone test at 21–24 days. If there is any reason to suspect that a cow has aborted, the oestrone sulphate test can be very useful in determining her status.

Mammary Development

Mammogenesis or development of the mammary gland occurs as a consequence of pregnancy and in the primiparous heifer changes can be detected as early as four months of gestation. In addition, a viscous brown secretion may be expressed from the teats. Of course, these changes are not apparent in the parous, lactating cow. Steroidal-type growth promoters can elicit identical changes in the mammary glands of heifers. This can be an additional confounding factor in countries and situations where they are used. It is only during the last few days of pregnancy when the udder becomes distended with colostrum that mammary development can be regarded as an accurate diagnosis of pregnancy

Ballottement

The abdomen of the pregnant animal begins to become distended from about seven months of gestation. If a hand is pushed firmly against the right side of the abdomen, the fetus may sometimes be felt to rebound against it. This technique is commonly used in cattle markets by prospective purchasers. However, it is not a reliable indicator of pregnancy particularly if the findings are negative.

CONCLUSIONS

A summary of available methods of pregnancy diagnosis is given in Table. Pregnancy diagnosis is a vital aid to reproduction management and the stock person should aim for the best combination of earliness and accuracy. For this reason, more than one diagnosis is recommended for each cow. As a minimum, an early milk progesterone test to achieve the earliest possible re-insemination of non-pregnant cows should be followed by a final confirmation of pregnancy, either manually or by ultrasound scan, at about 72 days, after which we consider the pregnancy to be relatively secure. Ideally, an intermediate diagnosis by ultrasound scan between days 28 and 42 should be carried out to get an early and accurate diagnosis. The cost should not be alarming when balanced against the cost of getting it wrong. The cost of the 'cow-side' progesterone test is very small when compared with the benefits, and the timing of the final diagnosis is not critical, so that larger batches of cows can be examined to reduce the cost. It should be emphasized that pregnancy diagnosis is of little use without a planned programme for treatment of cows that are diagnosed non-pregnant. For example, the exercise by Eddy (1983) described in Chapter combined early diagnosis of non-pregnancy using the milk progesterone test and the use of prostaglandin very effectively, as did the SAC procedure using scanning in heifers described above. The general cost benefit of the use of pregnancy diagnosis is very difficult to assess since many influencing factors vary from farm to farm. However, if regular pregnancy diagnosis is combined with efficient oestrus detection, a high standard of management and prompt action on non-pregnant cows, then it is likely to be extremely cost effective.

A summary of available methods of pregnancy diagnosis

Test	*Stage*	*Accuracy*	*Advantages*	*Disadvantages*	*Comments*
Non-return to service	3 weeks	About 50%	Early	Relies on good oestrus detection	Insemination at 'false oestrus' can lead to abortion
Milk progesterone	3 weeks	Positives 85% Negatives 100%	Early Non-invasive	False positives and later losses	Good test for non-pregnancy
bPAG/PSPB	4 weeks +	Fairly good	Relatively early	Usually needs blood May persist after loss	Can be useful, e.g. in heifers that are hard to palpate
Ultrasound scan	4 weeks +	Up to 100%	Accurate Relatively early	Cost of equipment	Can also detect problems. Recommended
Palpation	6 weeks(?) +	Up to 100%	High accuracy	Possible damage to cow or conceptus	Depends on operator. Early diagnoses especially risky
Oestrone sulphate	15 weeks +	Almost 100%	High accuracy Non-invasive	Only possible after 16–17 weeks	Useful confirmatory test
Mammary development	4 months (heifers)	Up to 100%	Non-invasive	Less useful in lactating cows	
Ballottement	30 weeks +	Positives 100%	Cost	Late	Little practical value in on-farm fertility management

REPRODUCTIVE PROBLEMS

Up to the present chapter we have concentrated largely on a description of reproductive function in 'normal' cattle. However, a number of problems can arise with the reproductive process so that normal function is lost. This can lead to a reduced level of reproductive efficiency or sub-fertility and, in some cases, to total failure of reproduction, or infertility. The most common reproductive problems of the bull and cow are discussed below, although some have been included in previous chapters where appropriate.

Reproductive Problems in the Bull

With the exception of bulls used for artificial insemination, bovine male fertility receives scant attention. However, the fertilizing ability of the bull is clearly critical in determining the reproductive performance of a herd, irrespective of the fertility status of the cows. Sub-fertility and infertility occur in bulls due to a variety of causes and the most common of these are discussed briefly below

CONGENITAL PROBLEMS

The Testicles

Testicular hypoplasia or defective testicular development can occur in all breeds and in an American survey was reported to occur in 1.3% of bulls (Carroll et al., 1963). Certain breeds seem to be more prone to the condition, which was very common in Swedish Highland bulls, occurring in up to 25% of animals. The testicles are noticeably smaller than normal, although this may be difficult to define with certainty due to a wide variation between 'normal' animals. The condition may be unilateral or bilateral and in the latter case the bull is likely to be infertile. In any event such bulls should not be used for breeding, because the condition may be of an hereditary nature. Cryptorchidism is the failure of one or both testicles to descend into the scrotal sac during fetal development. Although testosterone production is not affected, spermatogenesis is inhibited in affected testicles because they are maintained at body temperature. The condition is relatively rare in cattle (approximately 0.1% of bulls) but can be quite common in other species, e.g. the horse. An additional defect that occasionally occurs (approximately 0.2% of bulls) is the lack of development of segments of the male reproductive tract. An example of this is segmental aplasia or hypoplasia of the Wolffian duct from which the sperm transport system (which transports spermatozoa from the testicles to the urethra) develops. This condition is analogous to white heifer disease, discussed later in this chapter. Degeneration of the testicles may occasionally occur but is usually secondary to another disease process or possibly ageing. In this condition the testicles shrink in size and there is a loss of fertilizing ability.

Abnormalities of Spermatozoa

The assessment of normal semen characteristics is described in Chapter. A variety of morphological abnormalities of sperm occur in the bull and may be associated with infertility. These include detached sperm heads, acrosomal defects, coiled tails and immature sperm each containing a 'proximal droplet' at the junction between the head and midpiece of the sperm. These defects have been described in detail elsewhere (Arthur et al., 1996) and are considered briefly in Chapter. Poor quality semen is a major reason for bulls being declared unsuitable for AI purposes. Beef breeds appear to pose a particular problem in this respect.

The Penis and Prepuce

Prolapse of the prepuce is an eversion of the mucous membrane lining of the prepuce or sheath. It occurs occasionally and appears to be due to congenital hypoplasia of the retractor muscles of the sheath, a feature most common in polled bulls. If this problem is persistent it can lead to chronic infection of the prepuce. Deviation of the penis occurs in the bull and may be manifested in a variety of ways. For example, lateral, ventral and spiral deviations have all been reported (see Arthur et al., 1996). The causes vary but may often be congenital. For example, persistent attachment of the ventral surface of the penis to the inside of the sheath, a condition known as persistent frenulum, results in ventral deflection of the penis at erection or, in more serious cases, an inability to extrude the penis. Under normal conditions separation of the frenulum is complete by about nine months of age. Although these cases can be treated surgically, such bulls should not be used for pedigree breeding due to the possible hereditary nature of the condition. Benign virus-induced tumours of the penis and prepuce are common. These may grow quite large and cause irritation and bleed occasionally. The usual outcome is for the tumours to regress spontaneously, although surgical removal may become necessary. A relatively common traumatic injury is the rupture of the corpus cavernosum penis. The tough membrane surrounding the penile tissue ruptures, usually during service. The resultant haematoma or blood clot normally forms close to the sigmoid flexure. Following this, fibrous adhesions often develop between the penis and prepuce, making extrusion of the penis painful and difficult. Infections Infection of the reproductive tract can result in orchitis (inflammation of the testis), epididymitis or seminal vesiculitis. Organisms that have commonly been associated with these conditions are Brucella abortus and Corynebacterium pyogenes. The symptoms include pain in the inguinal area and elevated body temperature. Campylobacter (vibrio) fetus and Trichomonas fetus colonize the interior of the prepuce and hence infected bulls can transmit these organisms during natural service. However, these infections are rarely associated with lesions of the reproductive tract of the bull. Bovine herpes virus 1 is most commonly associated with infectious bovine rhinotracheitis (IBR) in young cattle but can also cause infection of the penis and prepuce (balanoposthitis) in adult bulls. Infection of the genital tract with mycoplasmosis is associated with low sperm motility in young bulls. Panangala et al. (1981) found that Mycoplasma bovigenitalium mixed with bull semen adhered to the spermatozoa and depressed their motility

Reproductive Problems in the Cow

As was discussed earlier, reproductive performance in European and North American dairy herds is declining steadily and is more than ever a highly important limiting factor in efficient herd performance. Problems in the female can affect the manifestation and detection of oestrus, the production, transport and fertilization of ova and the transport, implantation and survival of the conceptus. The underlying causes for these failures can be loosely divided into the categories of developmental, infectious and functional (including injury) problems. However, these distinctions are sometimes poorly demarcated, and tend to overlap. For example, retained placenta, a functional disorder, or injury caused by a difficult calving can predispose the reproductive tract to infection, and this can in turn lead to mechanical damage. The following is a brief summary of the more common clinical problems that result in reproductive failure. Detailed accounts of the causes, diagnosis and treatment of all reproductive problems in cows are covered in more specialized texts.

Developmental Problems

Most developmental problems in cattle, as in other species, result from genetic abnormalities, which can either be inherited from one or both parents or result from chromosome damage during oocyte development, fertilization and early embryo development. It is likely that a very high percentage of inherited abnormalities result in fertilization failure or death within a short time of conception, as will be discussed in the relevant chapters below. Occasionally, however, calves are born with inherited defects that may affect their survival or reproductive ability. Bovine leukocyte adhesion deficiency (BLAD) is an example of an inherited condition that normally results in the early death of homozygous carriers. It is now thought that even heterozygous carriers tend to have lower milk yields (Jánosa et al., 1999) and the possibility that reproductive performance may also be impaired may warrant investigation. For a review of BLAD see Gerardi (1996). A classic example of an inherited condition that affects reproduction is segmental aplasia of the Müllerian ducts, or white heifer disease. The name derives from the fact that the gene responsible for the condition is associated with white coat colour. The anomaly interferes with the normal development of the Müllerian ducts so that the reproductive tract becomes obstructed. The degree to which the various components of the reproductive tract are affected varies and depends on the time at which development was inhibited. The cow's ovaries are functional and therefore oestrus and ovulation may occur normally. Depending on the site of the obstruction, fertilization, conception and parturition may or may not be possible, since one uterine horn only may be affected. The condition was originally associated with the Shorthorn breed, but is still a recognized problem in currently used breeds, especially the Belgian Blue (Charlier et al., 1996). As was described in Chapter, the oviducts, uterus, cervix and anterior vagina develop from the embryonic Müllerian ducts. Normally, the ducts fuse to form a single cervix and vagina, but occasionally a double cervix is formed. Another developmental problem, freemartinism, occurs when a male and a female conceptus are present in the same uterus. It is a well-known phenomenon in cattle, in which there is a tendency for the placentae of twin fetuses to merge, but it is known to occur in other species, and is an increasing problem in sheep (Parkinson et al., 2001). If the placentae merge, the circulatory systems of the twins become interconnected. If one of the twins is a female, the development of the female sexual organs will be affected. This may be due in part to the effect of androgens or other hormones from the male circulation. Rota et al. (2002) measured plasma concentrations of antiMüllerian hormone (AMH) in bovine freemartins. Levels of AMH at birth were very high in males and in freemartins. They remained high in males for the first five months of life, but in the freemartins they rapidly decreased to normal female levels, showing that AMH came from the male twin. The freemartin female is likely to suffer from chimerism, a combination of cells of the two fetuses, and this could in itself be affecting the development of the female tract. Ennis et al. (1999) used Y-specific polymerase chain reaction (PCR) primer pairs to detect male cells in the blood of heifers born co-twins to bulls.

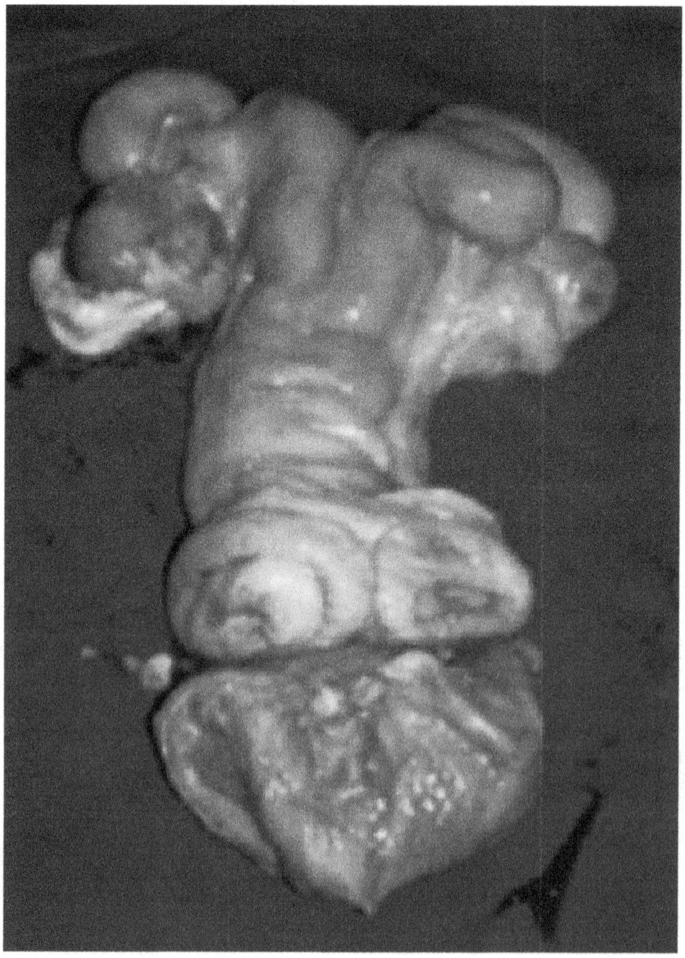

Reproductive tract from a heifer with a double cervix. Both were patent, leading into the uterus. In another heifer, the two fused to form a single passage into the uterus.

Between 1 and 2% of cattle births are twin births. Approximately half of these are heterosexual, i.e. 0.5–1.0%. Approximately 92% of heifers born as heterosexual twins exhibit the freemartin syndrome. The external genitalia of the freemartin usually appear normal, so that the condition may go unnoticed until it is time to breed the heifer involved. Heifers are affected to varying degrees: the more severe the condition, the more akin to their male counterparts will be the female gonads and reproductive tract. Unlike cows with white heifer disease, those with freemartinism are not likely to have functioning ovaries. It is probable that the condition would be more severe if placental fusion occurred relatively early in development. A typical freemartin reproductive tract is shown in Fig. and should be compared with the normal tract shown in Fig.. There is no cure for freemartinism, and it is important that the condition is diagnosed at an early age, so that no effort is wasted in rearing a nonproductive heifer. When female calves are purchased at a market, a potential buyer may use the 'pencil test', inserting a pen or pencil into the vagina to

Reproductive tract of a freemartin heifer. Note the underdeveloped uterus and very small ovaries.

determine its length. This may not detect small anatomical abnormalities that could still render the cow infertile. The test of Ennis et al. (1999) for the presence of male cells could be useful in practice. Satoh et al. (1997) gave pregnant mare serum gonadotrophin and human chorionic gonadotrophin to 4-month dairy heifers. Concentrations of oestradiol-17b and progesterone rose in normal heifers, but there was no response in freemartins. Again, we feel that this test may not detect less severely affected animals, which could nevertheless be infertile or subfertile.

REFERENCES

1. Amorim MR, Castilla EE, Orioli IM. Is there a familial link between Down's

syndrome and neural tube defects? Population and familial survey. BMJ 2004;328(7431):84

2. Anonymous. ACOG Committee Opinion #296:first-trimester screening for fetal aneuploidy. Obstet Gynecol 2004;104(1):215-7
3. Bach C, Torrent S, Cabrero D, Sabriá J. [First-trimester biochemial-ultrasound aneuploidy screening. Methodology and results]. [Spanish] Prog Obs Gin 2004;47:5-19.
4. Bahado-Singh RO, Sutton-Riley J. Biochemical screening for congenital defects. [Review] [71 refs] Obstet Gynecol Clin North Am 2004;31(4):857-72, xi
5. Bassett K, Lee PM, Green CJ, Mitchell L, Kazanjian A. Improving population health or the population itself? Health technology assessment and our genetic future. Int J Technol Assess Health Care 2004;20(2):106-14
6. Bersinger NA, Wunder D, Vanderlick F, Chanson A, Pescia G, Janecek P, Boillat E, Birkhauser MH. Maternal serum levels of placental proteins after in vitro fertilisation and their implications for prenatal screening. Prenat Diagn 2004;24(6):471-7
7. Bhide A, Thilaganathan B. What prenatal diagnosis should be offered in multiple pregnancy?. [Review] [37 refs] Best Pract Res Clin Obstet Gynaecol 2004;18(4):531-42
8. Bianchi DW. Circulating fetal DNA:its origin and diagnostic potential-a review. [Review] [80 refs] Placenta 2004;25 Suppl A:S93-S101
9. Bishop AJ, Marteau TM, Armstrong D, Chitty LS, Longworth L, Buxton MJ, Berlin C. Women and health care professionals' preferences for Down's Syndrome screening tests:a conjoint analysis study. BJOG 2004;111(8):775-9
10. Botto LD, Mulinare J, Yang Q, Liu Y, Erickson JD. Autosomal trisomy and maternal use of multivitamin supplements. Am J Med Genet. Part A 2004;125(2):113-6
11. Brajenovic-Milic B, Tislaric D, Zuvic-Butorac M, Bacic J, Petrovic O, Ristic S, Mimica M, Kapovic M. Elevated second-trimester free beta-hCG as an isolated finding and pregnancy outcomes. Fetal Diagn Ther 2004;19(6):483-7
12. Carrico A, Matias A, Areias JC. How important is a cardiac echogenic focus in a routine fetal examination?. Rev Port Cardiol 2004;23(3):459-61
13. Chen CP, Chern SR, Cheng SJ, Chang TY, Yeh LF, Lee CC, Pan CW, Wang W, Tzen CY. Second-trimester diagnosis of complete trisomy 9 associated with abnormal maternal serum screen results, open sacral spina bifida and congenital diaphragmatic hernia, and review of the literature. [Review] [31 refs] Prenat Diagn 2004;24(6):455-62
14. Chiu RW, Lo YM. Recent developments in fetal DNA in maternal plasma. [Review] [27 refs] Ann N Y Acad Sci 2004;1022:00:00 100-4
15. Chung JH, Yang JH, Song MJ, Cho JY, Lee YH, Park SY, Moon MJ, Lim HJ, Choi JS, Kim JO, Shin JS, Ahn HK, Han JY, Kim MY, Choi KH, Ryu HM. The distribution of fetal nuchal translucency thickness in normal Korean fetuses. J

Korean Med Sci 2004;19(1):32-6

16. Cicero S, Curcio P, Rembouskos G, Sonek J, Nicolaides KH. Maxillary length at 11-14 weeks of gestation in fetuses with trisomy 21. Ultrasound Obstet Gynecol 2004;24(1):19-22
17. Cole LA, Khanlian SA. Inappropriate management of women with persistent low hCG results. J Reprod Med 2004;49(6):423-32
18. Denis F, Ranger-Rogez S, Alain S, Mounier M, Debrock C, Wagner A, Delpeyroux C, Tabaste JL, Aubard Y, Preux PM. Screening of pregnant women for hepatitis B markers in a French Provincial University Hospital (Limoges) during 15 years. Eur J Epidemiol 2004;19(10):973-8
19. Dugoff L, Hobbins JC, Malone FD, Porter TF, Luthy D, Comstock CH, Hankins G, Berkowitz RL, Merkatz I, Craigo SD, Timor-Tritsch IE, Carr SR, Wolfe HM, Vidaver J, D'Alton ME. First-trimester maternal serum PAPP-A and free-beta subunit human chorionic gonadotropin concentrations and nuchal translucency are associated with obstetric complications:a population-based screening study (the FASTER Trial). Am J Obstet Gynecol 2004;191(4):1446-51
20. Egan JF, Benn PA, Zelop CM, Bolnick A, Gianferrari E, Borgida AF. Down syndrome births in the United States from 1989 to 2001. Am J Obstet Gynecol 2004;191(3):1044-8
21. Ekelin M, Crang-Svalenius E. Midwives' attitudes to and knowledge about a newly introduced foetal screening method. Scand J Caring Sci 2004;18(3):287-93
22. El Mouatassim S, Becker M, Kuzio S, Ronsin C, Gil S, Nouchy M, Druard L, Forestier F. Prenatal diagnosis of common aneuploidies using multiplex quantitative fluorescent polymerase chain reaction. Fetal Diagn Ther 2004;19(6):496-503
23. Elson J, Jurkovic D. Biochemistry in diagnosis and management of abnormal early pregnancy. [Review] [46 refs] Curr Opin Obstet Gynecol 2004;16(4):339-44
24. 24)Erickson JA, Ashwood ER, Gin CA. Evaluation of a dimeric inhibin-A assay for assessing fetal Down syndrome:establishment, comparison, and monitoring of median concentrations for normal pregnancies. Arch Pathol Lab Med 2004;128(4):415-20
25. 25)Fang Q, An N, Wu XY, Liu QL, Feng SH. [Short tandem repeat polymerase chain reaction used in prenatal diagnoses of the zygosity and common chromosomal trisomies in multiple pregnancies]. [Chinese] Chung-Hua i Hsueh Tsa Chih [Chinese Medical Journal] 2004;84(8):667-70
26. 26) Finberg HJ. An isolated echogenic heart focus is a low-level risk marker for Down syndrome.[comment]. J Ultrasound Med 2004;23(7):1008-9; author reply 1009
27. 27) Florio P, Imperatore A, Sanseverino F, Torricelli M, Reis FM, Lowry PJ, Petraglia F. The measurement of maternal plasma corticotropin-releasing factor (CRF) and CRF-binding protein improves the early prediction of preeclampsia. J Clin Endocrinol Metab 2004;89(9):4673-7

28. 28) Fong KW, Toi A, Hornberger LK, Keating SJ, Johnson JA. Detection of fetal structural abnormalities with US during early pregnancy. Radiographics 2004;24(1):157-74

29. 29) Goumy C, Bonnet-Dupeyron MN, Cherasse Y, Laurichesse H, Jaffray JY, Lacroute G, Geneix A, Lemery D, Vago P. Chorionic villus sampling (CVS) and fluorescence in situ hybridization (FISH) for a rapid first-trimester prenatal diagnosis. Prenat Diagn 2004;24(4):249-56

30. 30) Grijseels EW, Laudy JA, Galjaard RJ, Wildschut HI. [Prenatal investigations for Down's syndrome:medical-technical considerations and dilemmas arising from current screening methods]. [Review] [42 refs] [Dutch] Ned Tijdschr Geneeskd 2004;148(44):2166-71

31. 31) Gurbuz A, Karateke A, Mengulluoglu M, Gedikbasi A, Ozturkmen M, Kabaca C, Sahinoglu Z. Can serum HCG values be used in the differential diagnosis of pregnancy complicated by hypertension?. Hypertens Pregnancy 2004;23(1):01-Dec

32. 32) Gyselaers WJ, Vereecken AJ, Van Herck EJ, Straetmans DP, Martens GE, de Jonge ET, Ombelet WU, Nijhuis JG. Screening for trisomy 21 in Flanders:a 10 years review of 40.490 pregnancies screened by maternal serum. Eur J Obstet Gynecol Reprod Biol 2004;115(2):185-9

33. 33) Harris AH. The cost effectiveness of prenatal ultrasound screening for trisomy 21. Int J Technol Assess Health Care 2004;20(4):464-8

34. 34) Hauzman E, Fedorcsak P, Klinga K, Papp Z, Rabe T, Strowitzki T, Urbancsek J. Use of serum inhibin A and human chorionic gonadotropin measurements to predict the outcome of in vitro fertilization pregnancies. Fertil Steril 2004;81(1):66-72

35. 35) Higgins TN, Hanna AN, Hofer TL, Cembrowski GS. Measurement of inaccuracy and imprecision of HCG methods using dilutions of the WHO 4th IS-HCG standard and a pregnant patient's serum. Clin Biochem 2004;37(2):152-4

36. 36) Holmes A, Jauniaux E. Prospective study of parental choice for aneuploidy screening in assisted conception versus spontaneously conceived twins. Reprod Biomed Online 2004;8(2):243-5

37. 37) Hsieh LJ, Hsieh TC, Yeh GP, Lin MI, Chen M, Wang BB. Prenatal diagnosis of a fetus affected with Down syndrome and deletion 1p36 syndrome by fluorescence in situ hybridization and spectral karyotyping. Fetal Diagn Ther 2004;19(4):356-60

38. 38) Huggon IC, Turan O, Allan LD. Doppler assessment of cardiac function at 11-14 weeks' gestation in fetuses with normal and increased nuchal translucency. Ultrasound Obstet Gynecol 2004;24(4):390-8

39. 39) Hwa HL, Yen MF, Hsieh FJ, Ko TM, Chen TH. Evaluation of second trimester maternal serum screening for Down's Syndrome using the Spiegelhalter-Knill-Jones (S-KJ) approach. J Perinat Med 2004;32(5):407-12

40. 40) Jari SD, Fraer LM, Hogge WA. Association of undetectable unconjugated estriol on multiple marker screening with steroid sulfatase deficiency. Fetal

Diagn Ther 2004;19(1):43-8

41. 41) Kelekci S, Yazicioglu HF, Oguz S, Inan I, Yilmaz B, Sonmez S. Nasal bone measurement during the 1st trimester:is it useful?. Gynecol Obstet Invest 2004;58(2):91-5

42. 42) Kline J, Kinney A, Reuss ML, Kelly A, Levin B, Ferin M, Warburton D. Trisomic pregnancy and the oocyte pool. Hum Reprod 2004;19(7):1633-43

43. 43) Krantz D, Goetzl L, Simpson JL, Thom E, Zachary J, Hallahan TW, Silver R, Pergament E, Platt LD, Filkins K, Johnson A, Mahoney M, Hogge WA, Wilson RD, Mohide P, Hershey D, Wapner R, First Trimester Maternal Serum Biochemistry and Fetal Nuchal Translucency Screening (BUN) Study Group. Association of extreme first-trimester free human chorionic gonadotropin-beta, pregnancyassociated plasma protein A, and nuchal translucency with intrauterine growth restriction and other adverse pregnancy outcomes. Am J Obstet Gynecol 2004;191(4):1452-8

44. 44) Kuppermann M, Nease Jr RF, Gates E, Learman LA, Blumberg B, Gildengorin V, Washington AE. How do women of diverse backgrounds value prenatal testing outcomes?. Prenat Diagn 2004;24(6):424-9

45. 45) Lamont RF, Havutcu E, Salgia S, Adinkra P, Nicholl R. The association between isolated fetal echogenic cardiac foci on secondtrimester ultrasound scan and trisomy 21 in low-risk unselected women. Ultrasound Obstet Gynecol 2004;23(4):346-51

46. 46) Larrabee PB, Johnson KL, Pestova E, Lucas M, Wilber K, LeShane ES, Tantravahi U, Cowan JM, Bianchi DW. Microarray analysis of cell-free fetal DNA in amniotic fluid:a prenatal molecular karyotype. Am J Hum Genet 2004;75(3):485-91

47. 47) Lee MH, Ryu HM, Kim DJ, Lee BY, Cho EH, Yang JH, Kim MY, Han JY, Park SY. Rapid prenatal diagnosis of Down Syndrome using quantitative fluorescent PCR in uncultured amniocytes. J Korean Med Sci 2004;19(3):341-4

48. 48) Lim KI, Delisle MF, Austin SJ, Wilson RD. Cephalic index is not a useful sonographic marker for trisomy 21 and trisomy 18. Fetal Diagn Ther 2004;19(6):491-5

49. 49) Liou JD, Chu DC, Cheng PJ, Chang SD, Sun CF, Wu YC, Liou WY, Chiu DT. Human chromosome 21-specific DNA markers are useful in prenatal detection of Down syndrome. Ann Clin Lab Sci 2004;34(3):319-23

50. 50) Liu SS, Lee FK, Lee JL, Tsai MS, Cheong ML, She BQ, Chen SC. Pregnancy outcomes in unselected singleton pregnant women with an increased risk of first-trimester Down's syndrome. Acta Obstet Gynecol Scand 2004;83(12):1130-4

51. 51) Malone FD, Ball RH, Nyberg DA, Comstock CH, Saade G, Berkowitz RL, Dugoff L, Craigo SD, Carr SR, Wolfe HM, Tripp T, D'Alton ME, FASTER Research Consortium. First-trimester nasal bone evaluation for aneuploidy in the general population. Obstet Gynecol 2004;104(6):1222-8

52. 52) Markov D, Chernev T, Dimitrova V, Masneikova V, Leroy Y, Jacquemyn Y, Ramaekers P, Van Bulck B, Loquet P. [First trimester ultrasound screening

for structural and chromosomal anomalies in multiple pregnancy]. [Bulgarian] Akush Ginekol (Sofiia) 2004;43(5):11-Aug

53. 53) Markov D, Chernev T, Dimitrova V, Mazneikova V, Leroy Y, Jacquemyn Y, Ramaekers P, Van Bulck B, Loquet P. [Ultrasound screening and diagnosis of fetal structural abnormalities between 11- 14 gestational weeks]. [Bulgarian] Akush Ginekol (Sofiia) 2004;43(3):03-Oct
54. 54) Maslovitz S, Yaron Y, Fait G, Gull I, Wolman I, Jaffa A, Hartoov J. Feasibility of nuchal translucency in triplet pregnancies. J Ultrasound Med 2004;23(4):501-4
55. 55) Muller MA, Pajkrt E, Bleker OP, Bonsel GJ, Bilardo CM. Disappearance of enlarged nuchal translucency before 14 weeks' gestation:relationship with chromosomal abnormalities and pregnancy outcome. Ultrasound Obstet Gynecol 2004;24(2):169-74

CHAPTER 9

Livestock Production Systems

Chapter Contents

Livestock production systems are considered to be a subset of farming systems. A review of the literature revealed that most farming systems classifications are not backed by quantitative criteria, which would enable cases to be clearly allocated to one class. These classifications are closer to typologies. No attempts at developing a classification of world livestock systems by using quantitative statistical methodologies (cluster analysis and related methodologies) could be located in the literature. This probably relates to the lack of appropriate data sets for such approaches on a global scale.

INTRODUCTION

The word "livestock" is an umbrella term used for domesticated animals raised in an agricultural environment, with the intent of providing food, textiles, labor, or fertilizer to their owners. Common examples are horses, pigs, goats, cows, sheep, and poultry, although numerous other semi-wild animals including reindeer, yaks, camels, and emus could also be considered livestock. Humans have coexisted with domesticated animals for centuries, and the rise of farming and keeping animals probably contributed to a major shift in human culture.

The word can be taken to have several meanings, depending on interpretation. Livestock is sometimes referred to as "stock," in shorthand, reflecting the idea that the animals are property in addition to living beings. These animals are both living stock, or inventory, and the stock, or basis, of life for farmers and the people who rely on them. Raising animals is an important part of life for people all over the world.

Purely domesticated animals such as cows and horses are radically different than their wild counterparts. In some cases, the wild ancestors of domesticated livestock are actually extinct, because humans have selectively bred domesticated versions for so long. Domesticated livestock would probably have difficulty surviving in the wild, because it has been bred to be smaller and more docile than a wild animal would be. Semi-wild animals used as livestock, such as rabbits, have thriving wild populations in addition to domestic ones.

The uses for livestock are myriad. The most obvious is food, in the form of meat, dairy, and egg products. Few animals, however, are raised purely for their meat; the most notable exception to this is the pig, which is primarily a food animal. Most animals also contribute something else to the farm. Sheep, for example, have thick wool coats which are annually sheared to make textiles, and cows can provide physical labor as draft animals in addition to being a source of food. All livestock also produces plentiful amounts of manure, in the form of excrement, thus helping out in the farm garden as well.

The majority of livestock kept by smallholder farmers (small and large ruminants, pigs and poultry) are integral components of complex crop-livestock systems. Building a complete understanding of the biological, social and economic aspects of these systems is a necessary ingredient for determining intervention points and the development of appropriate technologies for smallholder farmers. The Livestock Production Systems program focuses on enabling equitable access to markets; optimizing crop-livestock systems and livestock production.

DIFFERENT TYPES OF LIVESTOCK

Larger farm animals with hooves, such as cows, goats, horses or pigs are part of a family known as "ungulates". Most ungulates are herbivores with plant-based diets, with the exception of pigs which are omnivores and therefore eat both meat and plants. Poultry, including turkeys, chickens and ducks are part of another important class of farm animal called "aves". In general, animals that are closely related zoologically require similar feeding and management practices.

The following paragraphs describe the major categories of farm animals:

- **Bovines – Big Ruminants:** include cattle and buffaloes. The term "ruminant" comes from the Latin ruminare, which means "to chew over again" as these animals regurgitate their food, chewing it once again to be able to digest it. Cattle are kept for both meat and milk, and are often seen as a sign of wealth and used as an alternative to cash.
- In North, Central and South America cattle dominate the ruminant populations. In Africa, the main concentrations of cattle are in the highland regions of Ethiopia, East and Southern Africa.
- India is the world's biggest milk producer, but industrialized countries are the most efficient in terms of yield per animal. Most buffaloes (97%) in the world are found in Asia. They are mainly kept for milk and tend not to be herded like cattle, but kept close to the village throughout the year. A major function of cattle and buffaloes in cultivated areas is to provide traction power. Pakistan and Southeast Asia have more buffaloes than cattle.
- **Small Ruminants:** consist of sheep and goats (or "shoats"). They can be kept in villages, but generally require a modest amount of land to support them - either for direct grazing or, in a limited number of situations, to provide fodder which is "cut and carried" to enclosed animals. Tethering a few sheep and/or goats around the house, near streams or along roadsides is a common practice.
- Shoats are also often kept alongside free-ranging cattle or camel herds. They can be kept in a wider range of environmental conditions than large ruminant species. Shoats are much more important than is often assumed - they are more common than cattle throughout a large part of the hotter, drier regions north of the tropics - in a band stretching from Spain and Libya to China.
- Goats are more common than sheep near the equator, and are known for being able to forage from even the most hostile, prickly plants. Goats are primarily found in Asia (63%) and Africa, while 39% of the world's sheep are found in Asia.
- **Monogastrics:** is a group that includes poultry (chickens, turkeys, ducks, geese, guinea fowl and pigeons) and pigs. The term refers to livestock that have only one stomach. They tend to have white meat and are generally easier to keep than larger animals. Many are left to forage for their own food within the smallholdings and are usually provided with protective housing at night. As human populations have increased and people have become more concentrated in villages and towns, they have started to rear more pigs or poultry (particularly

chickens). As a result, monogastrics are now found almost everywhere. The world poultry population is estimated to be 16 billion, which is fairly evenly distributed across regions, although Asia has the largest share. Asia also keeps 60% of the world's swine population, half of which are found in China.

- Although the above categories are the main focus of this module, other categories deserve mention as they are also very important:
- **Camelids:** include camels, dromedaries, llamas and alpacas. These are large animals with slender necks and long legs. They are "even-toed ungulates" that do not have hooves, but rather a two-toed foot with toenails and a soft footpad. Camelids are kept for milk (in the case of camels), meat, wool and transport. Of the 19 million camels in the world today, most are found in Africa. The large majority of smaller camelids such as llamas and alpacas are kept in South America.
- **Bees:** play an important role in pollination and in producing honey and beeswax, and help support small-scale farming around the world. There are over 20,000 known species of bees, although the actual number is probably higher. Bees are found on every continent except Antarctica, in every habitat that contains insect-pollinated flowering plants.
- **Fish:** are important to many small-scale farmers and fishermen. This module makes some mention of fish as part of small-scale cropping systems, such as mixed with rice, or small-scale (sweet-water) aquaculture systems.

Many other domesticated animals fall outside these categories. Some have small populations but can play very important functions in particular cultures and locations.

The same is true of rabbits which are found in many small-scale systems for food, whereas they are kept as pets in other cultures. In addition, what are sometimes known as "micro animals" (snails, locusts, rats, etc.) can be important in certain cultures and/or systems.

The Relative Importance of Different Livestock

Livestock populations are most densely concentrated in India, China and Europe – which are also the areas of high human population density. In general, livestock numbers - particularly of pigs and chickens - are increasing worldwide. However, in some places livestock numbers are falling, for example: pigs in parts of Europe; cattle in most of Europe; poultry in Scandinavia and Eastern Europe; and sheep and goats in central Europe and the former USSR. These declines are usually being offset by increases in productivity, so overall production levels in these regions remain more or less stable.

The biggest change happening to the world's livestock is a move away from traditional extensively-farmed poultry and pigs towards very intensive rearing systems that are more commercial and industrial. Nevertheless, ruminant species remain integrated in many farming systems within less developed regions, and their numbers are increasing. Small animals (shoats, poultry, etc.) continue to be critically important to the rural poor.

The patterns of global animal production have changed considerably over recent decades. For instance, the dynamic development of the pig and poultry industries in Asia and South America have meant that Europe's share of cattle, pig and poultry stocks and its contribution to global meat and egg production has decreased continuously since the 1970s. Nevertheless, European countries are still a major center of livestock and poultry husbandry with high technological, quality and safety standards.

Different animals are important in different contexts. In terms of meat production, poultry, cattle, shoats, and pigs are the most important. In terms of milk production, dairy cattle are by far the most important. From a poverty perspective, small animals such as poultry and shoats are the most numerous. From a dryland ecosystem perspective, people make use of camels and goats as the species best-suited to sustaining the environment as long as people practise sustainable herding strategies.

Different Functions of Livestock

They are four types of functions:

- Output
- Input
- Asset
- Socio-cultural

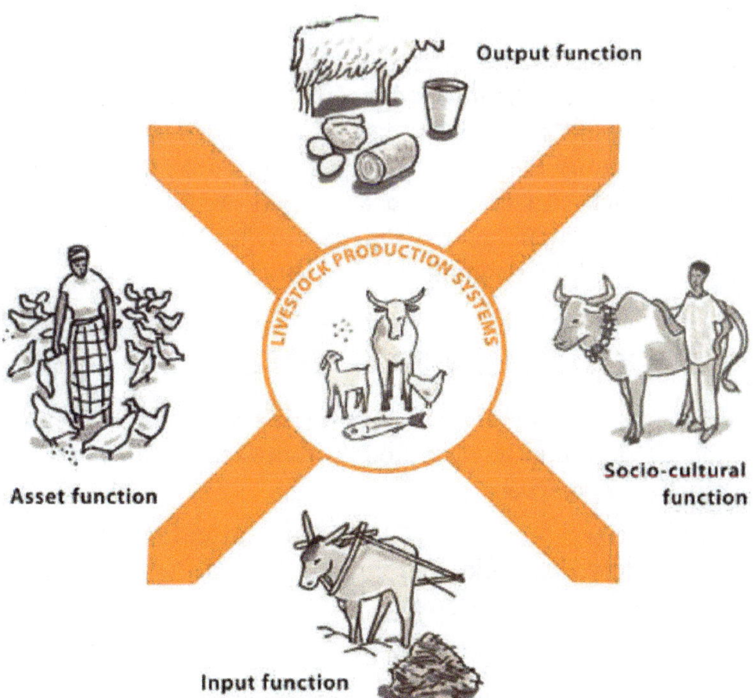

- **Output Function:** describes livestock's food and non-food products - such as milk, meat, wool, hair, eggs, honey, fat, fuel (manure/dung cakes) and medicinal products such as honey, urine, goat or camel milk. Meat, milk and eggs are the most important outputs, in terms of quantity, value and the number of animal species. Wool, felt, silk, fur and leather are important inputs into the clothing industry, while red dye, antlers, vellum and feathers all have specific uses.

- **Input function:** describes how livestock provide inputs for other agricultural sub-systems. In terms of crop production, livestock provide manure, urine and grazing of fallow land/stubble fields. Livestock contribute to the (re-) production of pastures, shrubs and trees by selectively removing biomass and thereby changing/controlling the composition of plant species, preventing bush fires, dispersing seeds (through their hair, hoofs, manure). Trampling can stimulate grass tillering, improve seed germination and break up hard soil crusts. Livestock also provide transport (draught power) for produce and people. The draught power of animals is harnessed in many, varied and changing ways. Animals are most often used in transport - pulling carts, sleds, etc. and in cultivation - pulling farm equipment such as ploughs, weeding devices – or for raising water from wells or threshing cereals. In some (particularly urban) areas free-ranging scavenging animals, such as pigs and goats, play a valuable role in garbage collection and sanitation. Besides the physical outputs mentioned above, herding communities provide services by using marginal areas sustainably for food production.
- **Asset function:** (or risk coverage) describes how people use animals as "capital" which can be sold to raise money in times of emergency (such as sicknesses or other crises).
- **Socio-cultural function:** describes how livestock relate to social status or fulfil cultural functions that express a certain "worldview", etc. Livestock are a part of the household and are indicators of social status. In many agricultural societies, a farmer's status (and particularly that of a pastoralist) depends on the number of livestock the family has. Many festivals and fairs are based on livestock and can include bullock cart racing, cock fighting, animal "beauty" contests, etc. Women often sing particular songs that express the valuable role that the animals play in their lives while cleaning, feeding or grazing and milking their animals.

LIVESTOCK PRODUCTION SYSTEMS

There are many different ways to classify livestock production systems: intensive or extensive, using modern or traditional methods, or producing for subsistence needs or for domestic or international markets, etc.

The classical categorization of the Food and Agriculture Organization (FAO) distinguishes between "solely livestock" and "mixed farming" systems, each of which is further subdivided into two other categories.

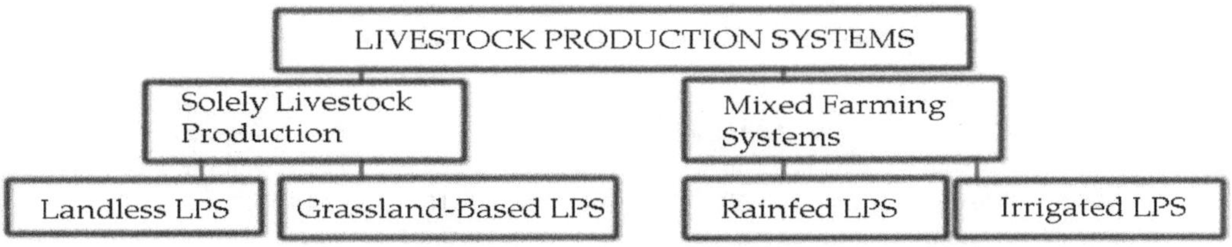

Solely Livestock Production Systems

The formal definition of these is that over 90% of the dry matter fed to animals comes from rangelands, pastures, annual forages and/or purchased feeds; and less than 10% of the total value of production comes from non-livestock farming activities.

Two types of systems can be distinguished here, as shown below.

Landless Livestock Production Systems

In these systems, less than 10% of the dry matter fed to animals is produced on the farm and the annual average stocking rates exceed ten livestock units (LU) per hectare of agricultural land. These systems are sometimes described as "landless" because the animals are physically separated from the land that supports them. However, since about 33% of the world's agricultural cropland is used to produce animal feed, the term "landless" is somewhat misleading. These systems produce about 45% of global meat output. The definition includes two types of systems:

- **Small-scale landless systems:** are typically found in urban and peri/urban areas and in rural areas with a high population density. These livestock keepers do not own croplands or have access to large communal grazing areas; they tether their animals, collect roadside grasses and by-products; or they can be "backyard" production systems. Keeping small animals contributes to their food security and provides a supplement to their incomes, while it is a relatively safe form of investment ("a bank on hooves").
- **Industrial landless systems:** are market-driven, intensive, specialized production systems. They are generally large-scale, but also exist at a small scale in peri/urban areas of developing countries. These livestock keepers purchase at least 90% of their feed from other enterprises. Industrial systems are common in Europe and North America and growing fast in pockets of South, East and Southeast Asia, Latin America and the Near East. They normally specialize on one single species (poultry, pigs or beef cattle) which are fed on feedstock (grain industrial and agricultural by-products) that is purchased from outside the farm. They contribute slightly more than two-thirds of the global production of poultry meat, slightly less than two-thirds of egg production and more than half of the world's output of pork, but are less significant in terms of ruminant production. The separation of industrialized livestock production from the land used to

produce feed also results in a large concentration of animal waste products, the disposal and storage of which can cause serious local environmental problems.

Grassland-Based or Grazing Systems

Grassland-based systems produce more than 10% of the dry-matter feed on the farm and have annual average stocking rates of less than ten LU per hectare of agricultural land. Such systems are the single largest form of land use and are currently estimated to occupy some 26% of the earth's ice-free land surface. They account for about 8% of global meat output. These are often subdivided into two different types of systems:

- **"Extensive" grazing systems:** cover most dryland areas of the world that are too marginal for crop production. Such areas tend to be sparsely populated and include the dry tropics and continental areas of southern Africa, central, eastern and western Asia, Australia and western north America. These systems are characterised by ruminants (e.g. cattle, sheep, goats) and camels browsing and grazing mainly trees, shrubs and grasses as well as other herbaceous plants. This often takes place on communal or open-access areas and largely in a mobile fashion (i.e., "pastoralism" or "semi-pastoralism"). Such systems produce about 7% of the total global beef supply, about 12% of shoat production and about 5% of global milk production.
- **"Intensive" grazing systems:** are found in temperate zones where high-quality grassland and fodder production can support larger numbers of animals. These areas tend to have medium to high human population densities and are mostly found in Europe, North and South America, parts of Oceania and some parts of the humid tropics. These systems are characterized by cattle (dairy and beef) and are mostly based on individual land ownership. They also include large-scale ranches in which farmers keep high-yielding cattle breeds on extensive grazing lands. They contribute 17% of the global beef and veal supply, about the same share of the shoat meat supply and 7% of global milk supply.

Mixed Farming Systems

In mixed farming systems, crop management and livestock rearing are interlinked activities. Mixed farming systems are defined as having more than 10% of the dry matter fed to animals coming from crop by-products or stubble; or where more than 10% of the total value of production comes from non-livestock farming activities.

Most of the world's ruminants are kept in mixed farms, which generally use outputs from one part of the farm to feed into other parts; i.e. the "input" function of animals mentioned above is important in these systems. Mixed systems account for about 46% of global meat output. These systems can be further subdivided according to their source and supply of water.

Rainfed mixed-farming systems

In these systems, more than 90% of the value of non-livestock farm production comes from rainfed land. These systems are found in temperate regions of Europe and the Americas and sub-humid regions of tropical Africa and Latin America. They are characterized by individual land / farm ownership and often make use of more than one species of livestock. They account for about 48% of global beef production, 53% of milk production and 33% of mutton production.

Irrigated mixed-farming Systems

In irrigated mixed systems more than 10% of the value of non-livestock farm production comes from irrigated land. These systems are most commonly found in East and Southeast Asia, mostly in areas with high population densities. Irrigated systems supply a large proportion of the world's supply of animal products, providing one-third of the world's pork, mutton and milk and one-fifth of its beef.

In general, the grazing and mixed livestock systems tend to be rather "closed", as outputs from one production activity (manure, crop residues) are used as resources or inputs to another. Intensive systems by contrast import and export large amounts of resources (feed, manure etc.) to and from other systems.

LIVESTOCK BREEDS

A breed is a specific group of domestic animals having homogeneous appearance (phenotype), homogeneous behavior, and/or other characteristics that distinguish it from other organisms of the same species and that were arrived at through selective breeding. Livestock breeding provides the population with various foods (milk, meat, lard, eggs, and so forth), and light industry with raw materials (such as wool, hides, and bristles). It also provides the tractive force of horses, oxen, asses, mules, camels, and deer. Other products of livestock breeding are organic fertilizer (manure), various kinds of fodder (skim milk, tankage, bone meal, and so forth), and medicinals (therapeutic serums and hormones).

Cattle Breeds

Common cattle breeds are:

Angus

Black Angus cattle, also called Aberdeen Angus, are the most popular breed in the U.S., and thanks to some excellent marketing, their meat is in demand, which means these cattle and crossbreds with mostly black markings often bring a premium at the sale barn.

This breed comes from northeastern Scotland, and was first brought to the U.S. by a Kansas rancher in 1873. When crossed with Texas longhorn cows, the hornless black calves brought winter hardiness to the mix. Angus are naturally polled (hornless), and have black skin and hair. They are moderately sized, generally good mothers, and are known for early development, ease of fleshing, good milk supply, and excellent marbling

Belted Galloway

Commonly called "Oreo cattle" because of their black color (possibly brown or red) with a white stripe through their middles, this breed started in Scotland as a solid-color cow, but got their belts through the introduction of Dutch Belted blood. They were first imported to the U.S. in 1950.

Although Belted Galloways are often purchased for their ornamental qualities, they do produce lean, quality beef. They're a medium-sized breed, but their carcass dressed weights can exceed 60% of their live weight. Belties have a double coat of hair, which allows them to keep warm in the winter without developing a layer of backfat like some other breeds.

Brahman

Brahman cattle come from India, and are the most common cattle breed in the world. Over the centuries, Brahmans have developed resistance to pests, parasites, and diseases, and the ability to survive inadequate food and harsh weather. They have a large hump over their shoulder and neck, upward-curving horns, large ears, and excess skin under their necks and chests, which helps keep them cool. They also are able to sweat better than most cattle, and secrete an oil which helps repel insects.

Charolais

The light-colored Charolais originated in France, where it was used for meat, milk, and drafting. The animals' large size and sturdy frame gave them the power to work in fields and pull wagons. The first Charolais came into the U.S. by way of Mexico in the 1930s. Because of a disease outbreak in Mexico, the breed was not allowed to be imported to North America until 1965. Therefore, many of today's American Charolais have other breeds in their lineage as well.

Charolais do well under a variety of environmental conditions. They graze aggressively in warm weather, withstand the cold, and have heavy calves. For this reason, adding a Charolais bull to a herd can improve the size and ruggedness of calves.

Dexter

Dexter cattle originated in southern Ireland, and came to the U.S. in the early 1900s. They are one of the smallest breeds of cattle, with full-grown bulls measuring 38 to 44 inches at the shoulder and weighing less than 1,000 pounds. Some have long legs and some short. Because of their size, they require less pasture and feed than larger breeds. They thrive in hot and cold climates, and are known for being gentle and easy to handle.

Dexters have a high rate of fertility and are easy calvers. They can be raised for both milk and meat. They can produce more milk for their weight than any other breed, and their milk yields up to a quart of cream per gallon. Their beef is slightly darker red than other breeds, and the small cuts are lean and graded choice.

Gelbvieh

This breed originated in Baravia, in southern Germany, and was originally developed for meat, milk, and work. It was introduced to the U.S. in 1971, through an artificial insemination program. Females are registered as purebred at 7/8 Gelbvieh, and bulls at 15/16. Bulls in Germany must undergo extensive tests to become A.I. sires.

Gelbviehs are red, with pigmented skin, and were originally horned. Due to breeding with polled foundation females in the U.S., though, many today are naturally polled. They are known for high fertility, ease of calving, being good mothers, and having quick-growing calves.

Hereford

The Hereford breed was developed in England in the 1700s to fulfill the expanding food market created by the industrial revolution. The original Herefords were bred for a high yield of beef and efficient production, and those characteristics are still important in the breed today. They were brought to the U.S. in 1817, and were useful for improving herds in the Southwest.

Because of their early maturity and fattening ability, Herefords became very popular in the U.S. As tastes changed in the 1950s, Herefords were bred to be leaner, with less fat and more red meat. Both horned and polled Herefords remain common in the U.S. They are known for their longevity, and for being docile, easy calvers, good milkers, and good mothers.

Holstein

Holsteins are best known as dairy cows, but those animals not used for breeding stock or milk production are raised for their value as beef cattle. Holsteins originated in Holland more than 2,000 years ago, and were brought to America in the 1850s as demand for milk grew in this country.

The black and white cattle are known for outstanding milk production, but their normal productive life span is only about six years. Healthy calves weigh 90 pounds or more, and mature cows reach 1,500 pounds.

Limousin

Limousin cattle may be as old as Europe itself; cattle in 20,000-year-old cave paintings in France are strikingly similar in appearance to today's breed. The golden-red cattle are native to France, and were used as draft animals to help turn rugged, rocky soil into fields for crops.

Limousins weren't imported into the U.S. until 1971, by way of Canada. Today, there are more than a million registered head here. In 2002, Lim-Flex, a pedigreed Limousin-Angus hybrid, was recognized.

Piedmontese

This Italian breed is a 25,000-year-old splice of two completely different breeds: the European Auroch and Pakistani Zebu. The breed was brought to North America in 1979.

Piedmontese are more muscular, disease resistant, and hardy than most beef cows. Due to a genetic abnormality, they are capable of developing muscle at an unrestricted rate, and with 14% higher muscle mass than most cattle, are considered double muscled. Piedmontese milk is also a primary ingredient in several Italian cheeses.

Red Angus

This breed was developed in Scotland in the 1700s, when large red English longhorn cattle were bred to native Black Angus cattle to produce animals heavy enough to be used as draft animals. One in four resulting calves were red. Both black and red offspring were initially considered purebred, but reds were banned from registration in 1917.

In the 1940s, American cattle producers started breeding reds cropped from the best Angus herds and formed their own breed, which aside from color, has the same features and benefits as Black Angus. Today, red Angus is the leading U.S. beef breed used in artificial insemination around the world.

Scottish Highland

This breed lived for centuries in the harsh, rugged Scottish Highlands, where it developed a resistance to many stress-related and other bovine diseases. It is among the oldest registered breeds.

Cold weather and snow have little effect on this breed, which has long hair rather than a layer of fat to keep it warm. This also makes for lean beef with little outside waste fat. They also do well in Southern climates, and will eat and thrive on brush and weeds other cattle pass by. Highlands have long horns, and long eyelashes and forelocks that protect their eyes from flying insects. They are considered to be even-tempered and intelligent.

Shorthorn

Shorthorns originated on the northeast coast of England, and were brought to America in 1783 and called Durham cattle. They were popular with settlers, since they were very adaptable, and could be used for meat and milk, and to power wagons and plows.

They can be either horned or naturally polled. Polled shorthorns were the first major beef breed to be developed in the U.S. in the 1880s. Both types of shorthorns are known for adaptability, mothering ability, reproductive performance, good disposition, longevity, and good feed conversion.

Simmental

This Swiss breed is among the oldest and most widely distributed in the world. They have been raised in the U.S. since the late 1800s, but their popularity waned until the late 1960s.

Most Simmentals are red and white, but there are no color restrictions on the breed. They are known for rapid growth development, milk production, and large size. Although primarily used as dairy cattle in Europe, American Simmentals are bred for beef production.

Texas Longhorn

This truly American cattle breed was shaped by a combination of natural selection and adaptation to the environment, stemming from the first cattle brought to North America more than 500 years ago. Due to a desire for more quickly maturing cattle, however, longhorns were nearly erased by crossbreeding by 1900. The breed was rescued from extinction and has regained popularity.

They are hard and adaptable, and are known for high fertility, easy calving, disease and parasite resistance, and longevity. Longhorns also eat coarse forage material more efficiently than most other breeds.

Watusi

Also known as African Ankole-Watusi, this breed traces its ancestry back more than 6,000 years, where long-horned domestic cattle were established in the Nile Valley. They are even pictured in Egyptian pyramid pictographs. Later, this giant-horned strain of cattle was owned by Tutsi kings and chiefs. Their horns, which can reach 12 feet in diameter, led them to become popular in European zoos.

These medium-sized animals have small calves, which makes Watusi bulls useful for breeding to first-calf heifers or other smaller breeds. They tolerate weather extremes, and do well in very hot climates. Their large horns actually cool them down by circulating blood, cooling it, and returning it to the body. Watusi cattle also produce low-fat, low-cholesterol beef.

LIVESTOCK BIOTECHNOLOGIES IN DEVELOPING COUNTRIES

Livestock contribute directly to livelihoods worldwide, providing not only food, but also non-food products, draught power and financial security. Livestock production already accounts for more than one third of the global agricultural GDP in developing countries, and this proportion is expected to increase. The rapidly increasing demand for livestock products, known as the "Livestock Revolution", has created opportunities for improving the welfare of at least some of the nearly one billion poor people who depend on livestock for their livelihoods. However, land degradation, environmental pollution, global warming, the erosion of animal genetic resources, water shortages and emerging diseases are all expected to present challenges to the growing global livestock sector.

Conventional technologies and biotechnologies in livestock have contributed immensely to increasing productivity, particularly in developed countries, and can help to alleviate poverty and hunger, reduce the threats of diseases and ensure environmental sustainability in developing countries. A wide range of biotechnologies are available and have already been used in developing countries in each of the three main sectors of animal sciences, which can be categorized as animal reproduction, genetics and breeding; animal nutrition and production; and animal health.

Animal Reproduction

In animal reproduction, genetics and breeding, artificial insemination (AI) has perhaps been the most widely applied animal biotechnology, particularly in combination with cryopreservation, allowing significant genetic improvement for productivity, as well as the global dissemination of selected male germplasm. Complementary technologies such as the monitoring of reproductive hormones, oestrus synchronization and semen sexing can improve the efficiency of AI. Embryo transfer provides the same opportunities for females, albeit on a much smaller scale and at a much greater price. Molecular DNA markers can also be used for genetic improvement, through marker-assisted selection using markers located beside genes of interest, as well as to characterize and conserve animal genetic resources. Use of most molecular marker systems depends on the polymerase chain reaction (PCR), which is an important technique for amplifying specific DNA sequences.

AI is practiced at some level in most developing countries. Its use is primarily in dairy cattle and peri-urban areas where complementary services including milk marketing are available. The high cost of liquid nitrogen for the cryopreservation of semen often restricts AI use far from cities. AI is usually used for crossbreeding with imported germplasm rather than for superior local genetics, due to the paucity of animal identification, recording and evaluation programmes. This lack of a system for the identification of superior animals precludes (along with lack of technical capacity) the use of more advanced technologies, such as embryo transfer or marker-assisted selection. Molecular biotechnologies in the area of animal reproduction, genetics and breeding have generally been limited to genetic characterization studies, usually through international cooperation.

Biotechnologies for animal nutrition and production are often based on the use of micro-organisms, including those produced through recombinant DNA technology. Fermentation technologies are used to produce nutrients (such as particular essential amino acids or complete proteins) or to improve the digestibility of animal feeds. Microbial cultures are used to increase the quality of silage or to improve digestion, when fed as probiotics. Recombinant bacteria have been developed to produce specific enzymes and hormones that improve nutrient utilization, which can increase productivity (e.g. somatotropin) and/or decrease environmental impact (e.g. phytase). Fiber-degrading enzymes have also been used to increase animal productivity and decrease environment pollutants.

Although data are scarce, amino acids and enzymes appear to be the most prominent and widespread nutrition-related biotechnology products used in developing countries. India and China have developed local industries to produce them. Various factors have limited the use of many other biotechnologies. For example, silage production is not common, thus precluding the use of microbial cultures. The uptake of recombinant somatotropin has been affected by low public acceptance, the lack of adequate and good quality feed and the low genetic potential of animals in developing countries. Fermentation of lignocellulosic materials to improve the quality of crop residues and forages has not been very effective.

Biotechnologies are used in animal health to increase the precision of disease diagnosis as well as for disease control and treatment. Monoclonal antibodies are used in immunology-based diagnostic methods, including enzyme-linked immunosorbent assays and radioimmunoassay. These methods may not allow the distinguishing of vaccinated from infected animals, so molecular biology approaches that detect specific DNA sequences are now preferred. Vaccination is also an indispensable method for maintaining animal health, and recombinant vaccines offer potential advantages over traditional vaccines in specificity, stability and safety. In addition, the sterile insect technique (SIT), usually applied as part of an area-wide integrated pest management approach, is used to improve livestock health within a certain geographical area by controlling insects that cause or transmit specific livestock diseases.

THE ROLE OF LIVESTOCK IN THE LIVELIHOODS OF POOR PEOPLE

Livelihoods involve the use of assets in activities to produce outputs to enable them to meet consumption requirements and aspirations and also to invest assets and activities for the future.

All this takes place in the context of an uncertain environment.

For many livelihood activities, production and income are irregular and intermittent, as, for example, seasonal cycles determine times of crop harvests, of livestock sales, and of opportunities for hiring out labor. There is also often a substantial degree of uncertainty about production and income, as they are affected by weather conditions, by crop and animal pest and disease attacks, by sicknesses and accidents (affecting the ability of family members to work), by market prices, and by changing policies and political influences (affecting taxes, subsidies, technical assistance, promotion of new technologies, security and political stability).

Irregular and uncertain patterns of production and income, however, do not generally fit with people's consumption and investment requirements. People have regular consumption requirements (for food and other daily needs), and they also face irregular investment and consumption needs (for example to pay for school fees, to buy animals or equipment, to construct buildings, to participate in annual festivals, or to participate in family or community social events such as births or weddings). There are also uncertain demands for expenditure to cope with accidents, sickness, or sudden demands from family members or others in the community and to take advantage of unexpected or unpredictable investment opportunities.

People therefore often face major challenges in matching the different production and income patterns on the one hand, with consumption and investment needs on the other. These challenges are particularly acute when people cannot access financial market mechanisms for saving, borrowing and insurance: in such situations they have to use different assets and activities to balance production and income patterns with consumption and investment needs. Recognizing the importance of this allows us to identify a number of different functions of assets and activities in people's livelihoods. Asset functions can be broadly classified according to contributions to production, savings, buffering, insurance, consumption and social integration. These contributions, which often overlap, are explained in more detail below by considering different functions of livestock keeping.

Productive assets and activities may make a variety of contributions to livelihoods. Regular cash income may be obtained from sales of products (milk, eggs, meat, hide & horn, wool & fibre, dung & manure), of services (work, transport, breeding sires, ceremonial purposes), or of live animals. These sales are likely to be more sporadic for poorer people with smaller numbers of animals, less intensive management, and weaker connections to markets. Sales of some livestock products and services are highly seasonal as a result of seasonal cycles in animal physiology (for example fertility, wool growth etc.), or as a result of seasonal demand (for example draught power and manure for crop production). There may also be seasonal market opportunities for specific products due to special food preferences for festivals. Many products and services that are sold are also used for domestic consumption (eggs, milk, meat, wool, and transport for example) or (as with draught power and manure) for production of other commodities which themselves may be consumed or sold.

Livestock keepers may use livestock as a means to fulfil a variety of savings type functions. Buffering (or consumption smoothing) describes the process where investments are made in livestock during periods when production or income exceeds consumption needs and then these investments are drawn upon later in the season when lower production and income are not sufficient to support consumption needs. Examples might be the purchase or breeding of an animal at harvest time and then feeding it with grain, to allow later sales income or meat consumption from the animal. Animals may also be kept explicitly to provide for some major expenditure (such as a major purchase or investment, or expenditure on school fees or an important social activity). This we term saving. In addition, animals may be kept just to provide some insurance against unexpected events that either reduce income or make extra expenditure demands (for example accidents or illness). A form of saving that is of particular interest is the process of accumulation – when the animal inventory of a household is built up over time as consumption needs and sales do not exceed the reproductive capacity of the livestock kept.

Livestock can sometimes also be used as collateral for borrowing. This is generally only possible with large animals. Borrowing may be needed to make up shortfalls in income at certain times of year (playing a buffering function), to meet unexpected consumption demands (playing an insurance function), or to make investments (supporting accumulation and production).

Finally, livestock may play important social integration functions in livestock keepers' society and culture. Ownership may confer status or animals may need to be exchanged or provided in certain traditional ceremonies or relationships which are important for people's position in local society.

Assets' Attributes

We now build on our understanding of livestock keeping functions in poor people's livelihoods by considering what features or attributes of different assets might make them more effective in fulfilling particular functions. Benefits from keeping and/or ownership of different types of livestock needs to be considered in terms of the fit between the attributes of livestock assets and their functions within the overall asset holdings and livelihood strategies of their owners.

Table 1 overleaf lists different attributes of assets and associated activities (these attributes being productivity, utility, security, holding costs, life, convertibility, complementarity and control), together with the asset functions to which these attributes relate, and the different components or elements of each attribute.

Table 1 should be largely self-explanatory but it may be worth noting that although it might be expected that high productivity and utility, low holding cost and longer life will be almost universally desirable asset attributes, what is desirable in terms of convertibility is more ambiguous and contextually determined. Theoretically, high convertibility would appear to give more flexibility and lower costs, but it is widely reported that poorer people (and less poor people as well) often impose disciplines and protection on their saving for longer term investment or for future consumption by placing savings 'out of easy reach' to prevent them from being cashed and spent by themselves or by others (for example relatives or neighbors) on more immediate consumption needs.

Table 1 Asset Attributes and Components

Main Attribute	*Contributing to which function(s)*	*Components*
Productivity	Production, income,	Productivity under expected, average or 'normal' conditions; variability; sensitivity to and resilience under different conditions; probability of these different conditions occurring ; appreciation of asset value
Utility	Income, consumption	'Normal' utility; variability; sensitivity to and resilience under different conditions; probability of these different conditions occurring
Security	All, especially saving	Risk of theft, loss of control or access; susceptibility to pathogens or other 'natural' event. For debts: risks to collateral or collateral substitutes
Holding costs	Detracts from all	Maintenance and input costs (including time, claims, etc) borne by different stakeholders: under normal conditions; variability under different conditions; probability of different conditions
		Depreciation in time and in use: under normal conditions; variability under different conditions; probability of different conditions
Life	All	Expected period over which asset will be held: under normal conditions; variability under different conditions; probability of different conditions
		Asset value profile (seasonal, lifecycle changes)
Convertibility	Sales income, savings, buffering, insurance	Exchange costs: under normal conditions; variability under different conditions; probability of different conditions
		Access: under normal conditions; variability under different conditions; probability of different conditions
		Lumpiness – related to unit value of sale and ease of sale
Complement-arity	Production, income	Effects on and of other assets and their functions
Ownership/ control	All	Private (individual, household); communal; public; gendered rights and responsibilities for disposal, acquisition, costs and returns

Livelihood Strategies

Having considered asset and activity functions in people's livelihoods, and the attributes affecting the way that assets and activities support these functions, we now need to consider the way that asset and activity functions and attributes relate to people's reliance on particular asset and activity mixes –their livelihood strategies. Considering asset and activity functions and attributes in the context of livelihood strategies is important for two reasons: first, if we can identify particular situations where different kinds of livelihood strategy may be more or less important, this should help us understand conditions under which different kinds of asset function and attribute will be important; second livelihood strategies draw attention to the importance of dynamic change when considering asset functions and attributes. We therefore describe first a simple classification of three types of livelihood strategy, and then examine the implications of this for the importance of different asset functions and attributes in different and changing situations.

We find it helpful to distinguish between three broad types of livelihood strategy, or three types of asset or activity contribution to livelihood strategies:

- 'Hanging In' where activities are engaged in to maintain livelihood levels, often in the face of adverse socio-economic circumstances
- 'Stepping Up' where current activities are engaged in, with investments to expand these activities, in order to increase production and income to improve livelihoods – with accumulation of productive livestock for example.
- 'Stepping Out' where existing activities are engaged in to accumulate assets which in time can then provide a base or 'launch pad' for moving into different activities that have initial investment requirements leading to higher and/or more stable returns – for example accumulation of livestock as savings which can then be sold to finance children's education, the purchase of vehicles or buildings (for transport or retail activities), migration, or social or political contacts and advancement.

Different people with different strategies will have different preferences as regards their asset function mix and as regards the importance of particular asset functions and of asset and activity attributes associated with these functions. We consider first which asset functions are likely to be important for livestock keeping in these different strategies, and then what determines which type of livelihood strategy is likely to be important in particular circumstances.

Within a 'hanging in' strategy livestock keeping will commonly contribute to four important functions (see table 2 overleaf): providing for subsistence consumption (through home consumption of meat, milk, eggs or fiber); supporting complementary (commonly cropping) activities (providing draft power and/ or manure); buffering against seasonality in income from other activities (for example cropping activities or seasonal labor); and providing some assets for insurance against unpredictable demands for cash.

Beyond these minimal maintenance functions, livestock keeping may enable advancement through accumulation either of more productive animals (the 'stepping up' strategy) or of a set of assets that hold values as savings to be used to 'buy in' to other assets needed to gain entry to other livelihood activities (the 'stepping out' strategy).

Table 2. Principle functions of livestock by livelihood strategy.

Livelihood strategy	Principle livestock functions
'Hanging in'	Subsistence Complementary production Buffering Insurance
'Stepping up'	Accumulation Complementary production Market production / income
'Stepping out'	Accumulation

What then determines which livelihood strategies (or combinations of strategies) will be adopted by poor livestock keepers? Livelihood strategies will normally be determined by the technical, institutional and market opportunities and constraints people face, and these in turn depend upon (a) peoples' access to assets and (b) upon the social, economic and natural environment in which people are located. This is illustrated in Table 3, which sets out situations where different livelihood strategies are likely to be more and less important, distinguishing between poor and less poor status and different market and technical (natural resource potential) situations.

Table 3: Likely livelihood strategies of poor and less poor livestock keepers, by market and natural resource potential

		Status	Local market opportunities	
			Low/ Stagnant	High/ dynamic
Natural Resource potential	Low	Poor	Hang in (very difficult – subsistence livestock?)	Hang in (more local non-farm based)
		Less poor	Step out (migrate)	Step out (local non-farm)
	High	Poor	Hang in (farm / subsistence?)	Hang in (farm and non-farm)
		Less poor	Step out (migrate) Step up ('exports')	Step out (local non-farm) Step up (local markets)

Whatever the market and natural resource potential of an area, very poor people are likely to give priority to 'hanging in' strategies as they struggle to maintain precarious and vulnerable livelihoods. However the focus of these 'hanging in' strategies is likely to vary, as regards the emphasis on different activities, with the agro-ecological (natural resource) and market opportunities in the area and with the particular assets they have (such as land, animals, skills, and social contacts). This is illustrated in table 3. Where natural resource or agro-ecological potential is low and the local economy (and hence market) is stagnant, conditions will be very difficult but livestock keeping may play a particularly important role in 'hanging in' strategies as a result of livestock's greater ability, as compared with crop based activities, to hold value as assets and to provide income at different times of year and under different seasonal conditions. Under conditions of higher natural resource potential crop farming may become more important to poor people's livelihoods (working either on their own farms or on the farms of others). However whatever the natural resource potential, under conditions of low market opportunity technological developments are unlikely to improve livelihoods by promoting increased production as without markets to dispose of incremental production, more production may have little value. Greater security, and more reliable (less risky) and faster accumulation may be more important goals, achieved through, for example, disease control, or more effective utilization of feed resources.

Where the local economy is more dynamic then there will be more local demand for unskilled labor and petty trading, and as these may offer income that is both less seasonally variable and less dependent on uncertain natural events than agricultural production, livestock may have less important buffering and insurance functions. On the other hand growing local markets may provide greater opportunities and higher prices in livestock sales, so productivity enhancing technological developments may have more to offer to some poor livestock keepers.

Table 3 also suggests how the predominant preferences of the less poor between 'stepping up' and 'stepping out', and the basis of these strategies, is likely to vary according to the potential for increased agricultural productivity (the Natural Resource potential) and for local markets. With stagnant local markets greater agricultural productivity will offer few livestock or crop based opportunities for 'stepping up' unless there are communications and linkages to support 'exports' to more distant markets.

The differing basis of the various strategies and the differences between them in more and less dynamic local economies together highlight the need to take account of the dynamics of change when analyzing asset and activity functions in the livelihoods of poor people. As noted earlier, poorer people who may be forced to give priority to 'hanging in' activities will nevertheless generally aspire and work towards stepping up and stepping out strategies. These strategies will change if people grow less poor (or conversely fall deeper into poverty). Strategies will also change with new social, technical and economic demands and opportunities as a result of economic growth, new infrastructure, new technologies, new services or new social norms in an area.

REFERNCES

1. "Livestock definition". Dictionary.com. Retrieved 23 November 2015.
2. "Cattle | Define Cattle at Dictionary.com". Dictionary.reference.com. Retrieved 2011-12-10.
3. "Agriculture: A Glossary of Terms, Programs, and Laws, 2005 Edition" (PDF). Archived from the original (PDF) on 2011-02-12. Retrieved 2011-12-10.
4. "Breeds of Livestock - Oklahoma State University". Ansi.okstate.edu. Retrieved 2011-12-10.
5. "RBST Gene Bank". Rare Breeds Survival Trust. Retrieved 29 December 2015.

6. Northern Daily Leader, 20 May 2010, Dogs mauled 30 sheep (and killed them), p.3, Rural Press
7. Simmons, Michael (2009-09-10). "Dogs seized for killing sheep - Local News - News - General - The Times". Victorharbortimes.com.au. Retrieved 2011-12-10.
8. "feed (agriculture) :: Antibiotics and other growth stimulants - Britannica Online Encyclopedia". Britannica.com. Retrieved 2011-12-10.
9. Markets from research to outcomes, Farming Matters, Challenge Program on Water and Food, June 2013
10. "2011 U.S. Greenhouse Gas Inventory Report | Climate Change - Greenhouse Gas Emissions | U.S. EPA". Epa.gov. 2006-06-28. Retrieved 2011-12-10.
11. http://www.worldwatch.org/node/6294
12. "Global warming breakthrough: way to stop cow gas - Unusual Tales - Specials". Smh.com.au. 2008-01-22. Retrieved 2011-12-10.
13. "Livestock's long shadow: environmental issues and options". Fao.org. Retrieved 2011-12-10.
14. "Sub-Saharan Africa news in brief: 10–22 April 2008". SciDev.Net. 2008-04-23. Retrieved 2011-12-10.
15. ^ Jump up to: a b "Global Methane Initiative | Global Methane Initiative". Globalmethane.org. 2010-10-0 Retrieved 2011-12-10.
16. "Cornell Science News: Livestock Production". News.cornell.edu. 1997-08-07. Retrieved 2011-12-10.
17. Pelletier Nathan and Peter Tyedmers (2011-01-17). "Forecasting Potential Global Environmental Costs of Livestock Production 2000–2050. Proceedings of the National Academy of Sciences of the United States of America 107.43 (2010): 18371-8374". Web of Science.
18. Starme, Elanor (2011-02-05). "LEVELING THE FIELD – ISSUE BRIEF #2 Environmental and Health Problems in Livestock Production: Pollution in the Food System" (PDF). American Journal of Public Health 94.10: 1703-709.
19. "FY-2005 Annual Report Manure and Byproduct Utilization - National Program 206" (PDF). United States Department of Agriculture.
20. "Risk assessment evaluation for concentrated animal feeding operations". United States Environmental Protection Agency.
21. "Animal agriculture: waste management practices" (PDF). United States General Accounting Office.
22. "The Facts". COWSPIRACY: The Sustainable Secret.
23. "Kangaroo farts could fight global warming | Courier Mail". News.com.au. 2007-12-05. Retrieved 2011-12-10.
24. Derner, Justin D., William K. Lauenroth, Paul Stapp, and David J. Augustine. "Livestock as Ecosystem Engineers for Grassland Bird Habitat in the Western Great Plains of North America." Rangeland Ecology & Management 62.2 (2009): 111-18. Web of Science. Web. Feb. 201

25. FAOSTAT. (Statistical database of the Food and Agriculture Organization of the United Nations.) http://faostat3.fao.org/
26. de Haan, C. D., H. Steinfeld, and H. Blackburn. 1997. Livestock & the environment: finding a balance. European Commission Directorate-General for Development.
27. Swanepoel, F., A. Stroebel and S. Moyo. (eds.) 2010. The role of livestock in developing communities: Enhancing multifunctionality. African Sun Media.
28. Fafchamps, M., C. Udry, and K. Czukas. 1998. Drought and saving in West Africa: are livestock a buffer stock? Journal of Development Economics 55(2): 273-305
29. Johannesen, A. B. and A. Skonhoft. 201 Livestock as insurance and social status: Evidence from reindeer herding in Norway. Environmental and Resource Economics, 48(4): 679-694.
30. Bell, L. W. and A. D. Moore. Integrated crop–livestock systems in Australian agriculture: Trends, drivers and implications. Agric. Systems 111: 1-12.
31. Kandulu, J. M., B. A. Bryan, D. King and J. D. Connor. 2012. Mitigating economic risk from climate variability in rain-fed agriculture through enterprise mix diversification. Ecological Economics 79: 105-112.
32. Asresie, A. and L. Zemedu. 2015. Contribution of livestock sector in Ethiopian economy: a review. Advances in Life Science and Technology 29: 79-90.
33. Bettencourt, E. M. V., M. Tilman, P. D. D. S. Henriques, V. Narciso, and M. L. D. S. Carvalho. 2013. The economic and sociocultural role of livestock in the wellbeing of rural communities of Timor-Leste. CEFAGE-UE, Universidade de Évora, Évora, Portugal.
34. Khan, N., A. Rehman and M. Salman. 2013. Impactul cre□terii animalelor asupra dezvoltării socio-economice în Nordul Indiei. Forum geografic 12(1): 75-80.
35. Ali, A. and M. A. Khan. A. 2013. Livestock ownership in ensuring rural household food security in Pakistan. J. Animal Plant Sci. 23(1), 313-318.
36. McSweeney, A. M and C. Raish. 2012. Social, cultural and economic aspects of livestock ranching on the Santa Fe and Carson National Forests. USDA Forest Service RMRS-GTR 276.
37. Gentner, B. J. and J. A. Tanaka. 2002. Classifying federal public land grazing permittees. J. Range Manage. 55 (1): 2-1
38. Torell, L. A., N. R. Rimbey, J. A. Tanaka, and S. A. Bailey. 200 The lack of a profit motive for ranching: implications for policy analysis. In: L. A. Torell, E. T. Bartlett and R. Larranaga (eds.) Current issues in rangeland economics. Proc. Symp. Western Coordinating Committee 55. N. M. State Univ. Res. Rep. 737.

Index

S

T

V

W